# Wings Across the Water

## Victoria's Flying Heritage 1871–1971



*Elwood White and Peter L. Smith*

HARBOUR PUBLISHING

Published by
Harbour Publishing Co. Ltd.
P.O. Box 219, Madeira Park, BC V0N 2H0
www.harbourpublishing.com

Page design by Peter L. Smith
Cover image by Graham Wragg
Cover design by Roger Handling
Photos and maps from the collection of Elwood White unless otherwise noted.

Printed and bound in Canada

Harbour Publishing acknowledges financial support from the Government of Canada through the Book Publishing Industry Development Program and the Canada Council for the Arts, and from the Province of British Columbia through the British Columbia Arts Council and the Book Publisher's Tax Credit through the Ministry of Provincial Revenue.

**Library and Archives Canada Cataloguing in Publication**

White, Elwood
Wings across the water: Victoria's flying heritage, 1871–1971/
Elwood White and Peter L. Smith.

Includes bibliographical references and index.
ISBN 1-55017-355-3

1. Aeronautics—British Columbia—Victoria—History. 2. Victoria (B.C.)—History. I. Smith, Peter L., 1933– II. Title.

TL523.W44 2005 629.13'09711'28 C2004-907465-2

# Table of Contents

## Preface by Peter L. Smith

Here is an unusual book, which should help to fill a major gap in British Columbia aeronautical history. Several able writers have recently explored specific aspects of aviation in the Victoria area—we think at once of Jack Schofield, Jim Brown, and the late Peter Corley-Smith—but there has been nothing even approaching a comprehensive history of flight on southern Vancouver Island. Because *Wings Across the Water* is a mainly pictorial survey, broad in scope, it does not presume to give definitive treatment to topics as complex, say, as the history of Patricia Bay/Victoria International Airport. But every major achievement from 1871 to 1971 is at least mentioned and documented in this one volume.

Almost forty years ago, Elwood White decided to produce a book along these lines. It was to be primarily a showcase for his enormous collection of historical photographs, many of which he had taken himself—going back to 1928, when he was ten years old. He realized that British Columbia's 1971 Centennial as a Province in Canada would coincide with the hundredth anniversary of B.C.'s first balloon ascension, a Victoria historic event. As an act of homage to his friend and fellow bus driver Frank Ellis, whose pioneer work *Canada's Flying Heritage* is still required reading, he thought he might call his book *Victoria's Flying Heritage, 1871–1971: the First 100 Years.*

A busy man with many interests, Elwood never found time to complete this aviation project, although he did draft many chapters of text. Along the way he co-authored a fine railway book and published a number of articles on transportation topics.

Victoria born and raised, Elwood White is recognized as an eminent authority on Vancouver Island trains, planes, steamships, buses, and anything else that moves. His transportation photos are in several public archives and a good many have appeared in print. Aeroplanes were his first love and have been a lifelong passion. He grew up on Shakespeare Street, close enough to Lansdowne Field to hear the roar of the Ford Trimotor engines. When classes were dismissed at Oaklands School, he would jog over to the airfield to see if there were any odd jobs available. As a teenager, he received his first plane ride from Lansdowne, courtesy of Cecil Eve, in return for sweeping out the B.C. Airways hangar. Not long after, Elwood White earned his own pilot's licence in Vancouver, at age 20.

Although I had known Elwood by reputation since I first read *Shays on the Switchback*, I never met him until he was over 80. With typical generosity, he had allowed me to reproduce a few of his precious railway photographs. Soon we were friends. One day he wondered aloud if I might be willing to help him publish a book that he had never managed to finish—a history of flying in Victoria. As he began to pull out box after box of stunning photographs, most of them never published, I realized that this was a work that had to be completed. Not only was it a unique documentation of aviation on southern Vancouver Island; it seemed to capture the whole story of human flight. Elwood could explain every fact and detail. His typewritten text was incomplete, but his command of the subject was profound. I told him I knew little about aviation, but would be eager to help.

Two years later, I rejoice in that decision. The man they call "Big El" is truly delightful; his warmth and humour enhance his prodigious knowledge and his razor-sharp memory. After leaving Oaklands, he soon found himself working for Canadian Pacific Steamships aboard the S.S. *Maquinna*. That may have been a school of hard knocks, but it obviously provided a first-class education.

He has been an excellent teacher. I was aware that there had been civic airports at Lansdowne and Gordon Head, but I had to be taught all of the fascinating details.

Elwood White's own notation on the photo: "Me and my gas job at the old Shelbourne Street airfield about 1937; Mount Tolmie is in the background."

Victoria's colourful seaplane history was news to me, though my first flight to Vancouver as a boy was from the Canadian Pacific wharf at View Royal. (I loved it, but was furious when my window blind was pulled down just before takeoff, as a wartime precaution—were they afraid of 11-year-old spies?) Now that I have an overview of local aviation from the balloon age to the jet age, I believe that our subject provides important insights into the social, economic, and political history of Victoria.

Elwood has also initiated me into the arcane jargon of aeronautics, and I appreciate its importance. I quiver with excitement in recognizing familiar registration letters. I can now toss these around like an air traffic controller. I may casually ask, "Do you mean Fairchild AJP or AKY?" But our book is not aimed primarily at aviation insiders, and there is a minimum of jargon.

As he and I worked out our plan, we decided on a pictorial format, with a documentary emphasis. It is not our purpose to dispense gossipy chitchat. Of course the book ought to be a good read, with a lively and interesting text, because many of these events are exciting and remarkable. Still, our most important goals are accuracy and precision. There is no need to include the dimensions and engine horsepower of every aircraft, because that generic information is readily available, and many enthusiasts will know it already. What is more important is to get specific local facts right, including significant names, dates, and places. On occasion, time-honoured errors and careless misinformation must be corrected. We are bound to have made mistakes ourselves, and we urge readers to correct us where we have gone wrong.

We hope you enjoy *Wings Across the Water* as much as we enjoyed preparing it. It was a colourful century, and Victorians can take pride in the city's aviation legacy.

Elwood White (age 20) hugs Fleet CF-ANF, the aircraft in which he earned his pilot's licence in 1938, through Foggin Flying Services at Sea Island Airport, Vancouver. This plane had earlier been flown by his two favourite Victoria "Musketeers," Hal Wilson and Maurice McGregor.

## EXPLANATORY REMARKS ON STYLISTIC CONVENTIONS

If our book is to be of any serious historical value, we must identify our factual sources. Footnotes can be distracting and hard to locate, but there are now accepted ways of avoiding them. The simplest solution is to provide a comprehensive bibliography at the back of the book, and to document a source by parenthetical reference in the text to author, year, and page number; for example, (Molson 1974: 225). If Molson wrote more than one book or article in 1974, the format is Molson 1974A or 1974B. Most readers will simply skip the parentheses or be satisfied in knowing that the source is K.M. Molson.

We have made extensive use of newspaper articles, mainly from the *Victoria Times* and the *Daily Colonist*, and it is important that these too should be documented. Newspapers are not always reliable historical sources for events that happened months or years before, but they are hard to beat for events that happened "today" or "yesterday." We owe a debt of thanks to those wonderful librarians who toiled for many decades to compile the *B.C. Newspaper Index*—a godsend. Our code for newspaper references is (year.month.day: page). An example might be (*Colonist* 1928.08.26: 1), a front-page article in the *Daily Colonist* on August 26, 1928, reporting that B.C. Airlines' Ford Trimotor has gone missing en route to Seattle. Like most of the civilized world, we were raised on the day/month/year convention, long favoured by Canadian aviation historians. However, a dating system such as ours avoids ambiguity, is computer-friendly, and appears now to be fairly standard.

Although we have no quarrel with the metric system, we often quote dimensions in imperial measure, so as to avoid creating strange anachronisms. If we know that a Patricia Bay runway was precisely 5,000 feet long, it would be contrived and misleading to describe it otherwise, and we respect our readers' ability to make mental conversions, if so desired.

## Acknowledgements

Many people have helped with this project, starting with Elwood's daughter Shannon, who typed countless pages of his manuscript more than thirty years ago. Warm thanks go to George and Nancy Maude, whose generosity and support have been exceptional. George is a peerless authority on the complex history of Victoria International Airport. After wartime service in the RCAF at the distinctive rank of Able Seaman (on a Pat Bay crash boat), he spent a long career with TCA /Air Canada. He has made a specialty of rescuing vintage aircraft.

John Howroyd, George Williamson, and Nils Christensen deserve special mention. Friends of Elwood's who have given recent help and encouragement include pilot and author Jack Schofield, former editor of *BC Aviator*; historian and archivist David N. Parker; and pilot Dan McIvor of Mars water bomber fame. Back in the 1960s and 1970s, Elwood made contact with many aviation pioneers now deceased, such as Frank Ellis, W. Harry Brown, Hal Wilson, Ted Cressy, Lloyd Jarman, Donald MacLaren, Sheldon Luck, and Gordon Ballentine.

We found new friends among those who came forward to help. It was a great privilege to meet Joyce McGregor, whose late husband Maurice was one of Victoria's most important gifts to aviation; she gave us full access to Maurice's collection of photographs, papers, and historic log books. Similarly generous were Donald G. Cameron, son of pioneer Victoria flyer Gordon A. Cameron; Dr. John A. Gray, son of Gordon Cameron's Aerial League partner Jimmy Gray; and Rear-Admiral William Hughes, nephew of Jimmy Gray and an avid family historian. After trying without luck to find a local relative of Alex B. Holden—another of our Victoria pioneers—we received a telephone call from Vancouver: his cousin Gloria (Holden) Simpson had heard of our project, and became particularly helpful. We had tea with Mrs. Eileen (Stubbs) Cox, age 97; she had flown on the Ford Trimotor with her husband John, office manager for Eve Brothers. Grant Olson lent us an amazing video of 1928–29 film clips owned by his grandfather, H.B. (Barney) Olson, who plays an important role in our story.

More help came from old friends: W.W. (Bill) Taylor, a glider pioneer and Victoria Flying Club member; Austin Smith, whose father Emerson Smith was a Canadian ace in the Royal Flying Corps; Nancy Newton, daughter of W. Hunter Wells; Claude Butler's niece Claudia Butler; Marilyn Woodward, who has two family links to our topic; and Harry M. Evans, an aviation enthusiast. Ken Wills and Eileen Wilson were quick to volunteer historic aeronautical material and memoirs of their father, *Victoria Times* newsman and City Alderman Archie Wills.

Our thanks go to Ian Baird in UVic Microforms, City of Victoria Archives, Saanich Archives, the British Columbia Archives, University of Victoria Archives, B.C. Aviation Museum at Victoria International Airport, Comox Air Force Museum, Canada's Aviation Hall of Fame, Department of National Defence, and B.C. Forest Service. We might have consulted original archival documents in Winnipeg and Ottawa, but chose to depend on secondary sources for some aspects of the task. In this electronic age, one can find almost anything on the Internet; we have been particularly impressed by the American website *Aerofiles*, but it is only one of many outstanding web resources.

The question of photo credits presents puzzling challenges. There is no problem with images purchased from the British Columbia and City of Victoria Archives, or those generously provided, for example, by the B.C. Aviation Museum, the *Times Colonist*, and UVic Archives. These all appear with standard credit lines. With similar ease, we can credit the very welcome photographs that we have received from kind friends, old and new.

The ambiguities arise within the huge Elwood White collection, upon which this book is mainly built. Perhaps a third of the total are photographs that he himself took over the years. Each of these that we use in the book is precisely labelled "Photo by Elwood White." Another large group consists of pictures that were memorable gifts or loaned negatives from friends; these appear, for example, as "Ted Cressy Collection," even though Elwood may have received them many decades ago.

The remainder, normally presented in our book without any credit line, must be interpreted simply as "Elwood White Collection." Though usually identified by date and subject, their source is now uncertain or unknown. The back of the photo provides no clue as to provenance. They usually appear to be prints from original negatives, and may have been in the collection for as long as fifty or sixty years. We believe earnestly in the value of precise photo credits, but there is no way to pin a label on pictures in this category. That is unfortunate, since these are all old and often historic photographs, and it would be at least a courtesy to acknowledge their source. We apologize to anyone who may recognize one of these unattributed images from the distant past.

# PART ONE

# AERIAL PIONEERS

## 1871–1913

B.C. Archives E-06756

When the Province of British Columbia launched its Centennial Celebration in 1971, the year began with an act of homage (see above) to the historic first balloon ascent one hundred years before. It is astonishing to reflect on the progress of aviation in B.C.'s first century. The summer of 1971 witnessed the delivery to Air Canada of the first Boeing 747 jumbo jets. On southern Vancouver Island, Victoria International Airport had become one of Canada's major air terminals, only thirty-odd years since it was developed in wartime by the Royal Canadian Air Force.

---

In this book, we shall be reviewing and exploring the course of aviation in British Columbia's Capital Region during those first hundred years from 1871 to 1971. Our focus is partly technical, since we wish to place on record as precisely as possible the developments in aviation that occurred in Victoria during that period, including the specific aircraft and the pilots who flew them. Wherever possible, however, we shall try to give the nuts and bolts of the story a broader human dimension. Our narrative will often touch on events of wide general interest, and it can be shown that the history of local aviation is closely connected with the social, economic, and political history of southern Vancouver Island. One recurrent theme, for example, is the problem of achieving visionary and unified civic decision-making in a small community that is an odd patchwork of independent municipalities.

This first section, entitled "Aerial Pioneers," is essentially a preamble to the main account, since there was not much happening in local aviation before the First World War. That can be said of almost any community in Canada. The progressive development of hot-air balloons, dirigible airships, and primitive heavier-than-air machines is a stirring and romantic story, and each stage was essential as a prelude to the next. However, it was only with World War I—the bloody and devastating "Great War"—that aviation came of age, pointing the way to many practical applications in the 1920s and 1930s.

Still, we know that Victoria was the scene of some colourful and interesting pioneer activity in Canadian aviation. By far the most significant achievements were those of William Wallace Gibson, a Victoria resident whose inventive genius has already been well acknowledged locally and nationally. The other events described in Part One reflect how aviation might have touched any community such as Victoria in the decades leading up to World War I. Generic though they may be, these events are inevitably coloured and flavoured by their local context.

## 1. Canada's New Province Gets a Flying Start ▪ July 20, 1871

### A Balloon Rises over a James Bay Picnic Celebration

The event was of no cosmic significance, but the timing was neatly symbolic. On Thursday, July 20, 1871, as British Columbia was celebrating its new status as a Province of the Dominion of Canada, Victorians can be said to have witnessed the birth of aviation in their city. The Provincial Capital observed the political occasion with a genteel extended picnic, which had as its climax a sensational balloon ascent (or "baloon," if we use the preferred spelling of the *Daily Standard*).

To judge from the *British Daily Colonist* report (July 22, 1871), this event might have been a bizarre tragedy:

> During the evening skyrockets were fired and blue lights burned, and a large balloon with a small boy in it, made (we mean the balloon) by Mr. Howse of the Lands and Works, was sent up at ten o'clock. The balloon was watched till it was out of sight and has not since been heard of.

We can rest assured, however, that the child was some kind of dummy, for the balloon ascent was definitely unmanned (and un-boyed). The *Daily Standard* describes the entire picnic at great length in its Saturday edition (July 22, 1871, p. 3), concluding as follows:

> About nine o'clock the fireworks were announced. Mr. Roscoe, an energetic member of the Committee of the Institute, superintended the display of pyrotechnics. These, with the baloon [*sic*] ascent under Mr. Howse of the Lands and Works office, were very successfully carried through.
>
> At 10 o'clock rose the monster baloon. Mr. Howse's efforts proved entirely successful. The aerial navigator ascended perpendicularly to a great height, then proceeded northwards and over the city, keeping in sight for about ten minutes, all the while being watched by the large concourse of gazers. An attempt to send up two smaller baloons, which were procured with some fireworks from Olympia, was not so fortunate. They refused to go skywards, and were deferred for another trial. A few more ships' rockets, and the crowd began by this time to grow thin, when at midnight most of the lamps and lanterns were removed, and all was again as still as midnight should be.

CONFEDERATION DAY.

THE MECHANICS'

INSTITUTE

WILL HOLD A

PICNIC

—AT—

MEDANA'S GROVE,

On THURSDAY, 20th July, 1871, where the day will be celebrated by a variety of Amusements in honor of the occasion.

PROGRAMME.

Salute at 1 P. M,
Address on "CONFEDERATION" at 3 P. M
Dancing throughout the Day and Evening,
Foot races and other Sports,
An Efficient Band will be in attendance,
A Renowned Fortune Teller is also expected to be in attendance,
Grand Display of Fireworks at 9 P. M.,
Baloon Ascent (the largest ever seen in the Province) at 10 P. M.,
The Piano will be Raffled at 6 o'clock if the Tickets are disposed of. Tickets for sale on the grounds.
Admission $1, Boys 50cts., Ladies Free.

Tickets to be had at Hibben & Co's, or from any Member of the Committee.

Mrs. McDonell will be on the grounds with Refreshments.

Mr. Moss will also be on the grounds with Confectionery, Fruit, etc.

jy19 EDWIN JOHNSTON, Hon. Sec.

Victoria Daily Standard 1871.07.19 2

By 1871, balloon ascents—even manned ascents—were old hat in North America, and British Columbia may seem to have been way behind the times. But as events would unfold over the next one hundred years, Victoria was to play a significant role in the history of Canadian aviation. Because of its island location, its residents came early to recognize the potential importance of air transport, and the city was associated with a number of pioneer endeavours. In the fullness of time, the City developed a major international airport. Even this modest 1871 balloon ascent, we can note, was undertaken and carried out as a local initiative. Many other notable achievements during our first century of aviation were similarly "made in Victoria."

## 2. A Procession of Visiting Hot-Air Professors ▪ 1880–1908

Nanaimo appears to have been the first city in British Columbia to experience a manned balloon ascension. The event occurred on the evening of June 9, 1880, and the hero of the hour was an acrobatic aeronaut identified only as Professor LeClaire. Western Canada was not in the vanguard of hot-air aviation: as early as 1862, Professor M. Ayers had been performing similar feats in Quebec and Ontario (Fuller 1983: 8). For whatever reason, this breed of daring exhibitionist had been consistently granted professorial status; and Victoria would have its share of visiting professors before the advent of heavier-than-air flight. At the end of this period, one of the most enterprising balloonists to perform in the city was actually a Victoria resident. Because he was also an inventor, he will appear in the next chapter.

**NANAIMO—June 9, 1880**
**PROFESSOR LeCLAIRE**
**Baloon [*sic*] ascension.**

On Wednesday evening [June 9], Prof. LeClaire made an ascension on his monster Baloon from the show grounds on Skinner Street. The baloon was inflated by an improvised furnace with smoke and heated air. Instead of the usual basket or box at the bottom of the baloon, there was a single trapeze. On letting go, the baloon shot up almost perpendicularly for close on 2000 feet, with Prof. LeClaire sitting on the trapeze quite unconcerned. The upper currents gradually wafted the baloon in the direction of the V.C.C. [Vancouver Coal Company] wharves, and the air ship struck the water close to the ship lying at anchor in the harbour. A boat put out and took the daring aeronaut on board, at the same time securing the baloon. The Professor returned to the show, changed his dripping clothes and the entertainment commenced. The tent was full, to witness the varied performance of the atheletes [*sic*], the educated pig, etc. No ascension took place on Thursday, owing to the slim attendance. The company performed at Wellington last night.

—*Nanaimo Free Press* 1880.06.12: 3

In reviewing this sequence of balloon flights, we shall transcribe verbatim a number of contemporary newspaper accounts. These tend to be very vivid, evoking the spirit of the period and needing little additional comment.

All Vancouver Islanders can take pride in the pioneer role played by Nanaimo in this colourful sphere of aviation. The rest of our manned balloon ascents all took place within Greater Victoria. The earliest was at Esquimalt Harbour, near Signal Hill. The others were usually presented on the Oak Bay waterfront, though the last on record (1908) was staged at Willows Fairgrounds.

**ESQUIMALT—July 27, 1891**
**PROFESSOR WOODHALL**
**Many witnessed it.**

A crowd of 800 or more witnessed Prof. Woodhall's balloon ascension at Esquimalt, yesterday, and, though the trip of the aeronaut was not quite satisfactory, it pleased the audience immensely. The inflation of the big bag commenced at 5 o'clock, and half an hour later the balloon rose to a height of 1,000 or 1,200 feet, and then dropped on the rocky ground, near the canteen. There was no chance for a parachute descent, and this feature is reserved for a second exhibition, which will probably be given on Saturday next.

—*Victoria Daily Colonist* 1891.07.28: 1

Apparently there was no parachute descent, a recent technical innovation that was still very dangerous and uncommon. One such was staged in Victoria five years later by our next visiting professor.

**OAK BAY—August 26, 1896**
**PROFESSOR FRANK MILLER (1st Visit)**
*ADVANCE ANTICIPATION*

Ballooning, which not many years ago was looked upon by the general public as a variety of circus performance and nothing more, has now established itself as an important branch of navigation worthy of more than the passing attention of scientists. The balloon and the parachute already have their recognized place in military operations, while Prof. Andree's project of making his way to the long sought pole by balloon has greatly interested the entire civilized world in the possibility of long distance aerial cruising.
(*continued next page*)

(*continued from page 4*)

Victorians have not, strange to say, had the pleasure of witnessing a balloon ascension for more than ten years, while a parachute jump has never yet been made in the near neighborhood of the city. For these reasons as much as because of the novelty, there will undoubtedly be an immense crowd at Oak Bay this afternoon to see the unique entertainment provided by Prof. Frank Miller. He makes his ascension from in front of the Mount Baker hotel at 4 o'clock, and being a daring as well as an experienced aeronaut guarantees to attain an altitude of at least 3,000 feet—he may reach 7,000 feet—before entrusting himself to the parachute. Every detail of the exhibition has been carefully attended to, and for the accomodation [*sic*] of the public a ten-minute tramcar service has been promised. The sight is one which there are few opportunities to witness, and the exhibition is entirely free to the public.

—*Victoria Daily Colonist* 1896.08.26: 5

**OAK BAY—August 26, 1896**
**PROFESSOR FRANK MILLER (1st Visit)**
*THE ACTUAL EVENT*
**THE RISE AND FALL OF MILLER**

Any person disputing the accuracy of Victoria's census would have had all doubts dispelled had he been at Oak Bay beach yesterday afternoon when Professor Frank Miller made his balloon ascension and parachute drop, concluding the performance with a bath in the over-cool waters of the Straits. The latter feature was not advertised but was, like the spectacle proper, entertainment for the public free, gratis, for nothing and without charge.

This may have had something to do with the large attendance, but in any event the "Professor" was honored with the attention of the largest open air gathering that private enterprise has yet brought together in Victoria. They came on foot and by bicycle, in tramcars, hacks, carriages, boats and baby wagons, but they all managed to be on hand by four o'clock, with the ladies and children in the majority. This was the hour set for the ascension, but owing to the unfavourable wind it was almost six [p.m.] when Miller called out to "let her go," and the big grey gas bag shot upward, the "professor" below describing a series of pendulum movements with the lurching of the air ship, and the crowd expressing its several feelings in a choice of selection of "Ohs" and "Ahs."

The wind was from the land and the balloon swept at once seaward, rising with unusual rapidity and careening considerably when a height had been reached at which the aeronaut resembled a small black kitten suspended by its tail.

It was at this juncture that the parachute was cut loose and the people held their united breath to see if it would open properly. It did, and worked very well in the descent, which was, however, a rapid one. The big umbrella fell near the little rock island at the entrance of the bay, Miller jumping clear when about twenty feet from the surface. His object was to avoid the island, but unluckily he struck an outlying spur of rock and his ankle was painfully cut in consequence. It was a short swim to the island and there he and the parachute were picked up a few minutes later by A. Sarantis and D. Seferles with a row boat. The balloon fell and was recovered quite a distance further out by C.T.W. Piper.

On the whole the performance was a very satisfactory one and fully merited the generous applause accorded. The consensus of opinion seemed to be that the life of the aeronaut like that of the policeman is not a happy one, and he fully deserved the glad hand as well as to wear the prefix "Professor" just as long as it pleased him.

Thanks very largely to the enterprise of Mr. Virtue in providing reasonable entertainment for the public, Oak Bay has this summer firmly established itself as the most popular near-by resort to which Victorians and visitors repair for rest, recreation and amusement. The twice-a-week band concerts—one of which was given last evening—have done much toward making the people acquainted with the beauties and attractions of the Bay, while such exhibitions as that of yesterday are sure to draw the crowds. Professor Miller repeats his unique performance by special request on Saturday evening at six, afterwards going to the Mainland for the carnival.

—*Victoria Daily Colonist* 1896.08.27: 5

The final paragraph answers a puzzling question: why is this wonderful entertainment being provided free of charge? (That was surely not the normal practice of itinerant balloonists!) Here we discover an entrepreneurial angel who had a vested interest in promoting the merits of Oak Bay. This was Mr. John Virtue, Manager of the grand Mount Baker Hotel (1893–1902), which faced the balloon ascension site across Mt. Baker Avenue (now Beach Drive). We may conclude that the balloon played some role, however small, in the public discovery of that very attractive part of the city, a district that became incorporated exactly ten years later, in 1906.

**OAK BAY—May 18 and May 23, 1901**
**PROFESSOR EARLSTON**
**BALLOON ASCENDED**
**But Without Prof. Earlston,**
**the Daring Aeronaut.**

Between three and four thousand people went out to Oak Bay yesterday afternoon to see the balloon ascension and parachute drop. They returned to the city, bitterly disappointed, for only a part of the proposed programme was carried out. The parachute drop did not come off; the balloon ascended, but without "Professor" Earlston. The "Professor" and his assistants inflated the balloon on an open space north of the Mount Baker hotel. East of this spot a telephone wire runs to the hotel. This wire was the cause of the fizzle.

—*Colonist* 1901.05.19: 2

Seeing that his trapeze was going to strike the wire, Earlston let go his hold as the balloon began to rise; it sailed away without him, dropping into the sea east of Mary Tod Island (a.k.a. Jimmy Chicken Island).

That aborted flight occurred on the afternoon of May 18. On May 23, Earlston tried a second time, achieving a successful ascent to 2,800 feet, with a parachute drop.

**OAK BAY—August 7, 1902**
**PROFESSOR FRANK MILLER (2nd Visit)**
**Great Crowd at Oak Bay to See Prof. Miller.**

The capacity of the Oak Bay street car service was taxed to handle the crowds who went out to that beautiful beach resort to witness the parachute jump of Professor Miller, while his balloon was high in air. The evening was an ideal one, there being scarcely sufficient breeze moving to cause a ripple. The inflating of the huge hot-air bag, as usual, afforded infinite amusement for the small boy and many of his seniors who lent willing hands in assisting the aeronaut during the process of preparing his balloon.

—*Colonist* 1902.08.08: 2

This successful flight was very much like Miller's 1896 ascent and parachute jump, except that he had a soft landing in the water about a quarter mile from shore. Miller had jokingly promised to drop a live chicken (*sans* parachute) at the zenith of his flight. His supposed intent was either to placate or get revenge on Jimmy Chicken Island, where he had been injured six years before. When the fowl deed was not committed, the youngsters in the crowd were reported to be keenly disappointed.

Courtesy Victoria Times Colonist

Willows Exhibition Grounds in the late nineteenth century. This exhibition building, an ornate Victorian legacy, was destroyed by fire in 1907, a year before Frank Sylvan made his balloon ascensions.

In 1887, the City of Victoria purchased from the John Tod estate a 65.3-acre parcel of land that became Willows Exhibition Grounds (or Fairgrounds). For many years this was the site of the annual Victoria Exhibition, a traditional agricultural fair. Nearby, at Fort Street and Epworth, the famous Patrick Arena (1911–1929) witnessed the first hockey game on artificial ice, and also the Victoria Cougars' victory in the 1925 Stanley Cup series. The Barney Olson Arena (1941–1944) was a horse show building converted to an ice rink.

The City continued to own the Fairgrounds until the land was subdivided for housing at the end of the 1940s. Until then it provided Victoria's main horse-racing track and grandstand, a venue for various forms of public entertainment. Willows first hosted aerial events in the six years before World War I. Then the Fairgrounds became Willows Camp, a large military training base. For a few years after the Great War, the racetrack area was identified as Willows Aerodrome.

Even though Victoria's first manned balloon ascent for which we have documentary evidence took place in Esquimalt, on July 27, 1891, the favoured locale for these spectacles was on the east side of Greater Victoria, in what became in 1906 the Municipality of Oak Bay. Apart from seaplanes in the Inner Harbour, almost all the historic events in Victoria aviation occurred outside the modern city limits—in Oak Bay, Saanich, or Esquimalt, and later in North Saanich and Sidney (Patricia Bay Airport). For historic reasons, however, many of the earlier activities came under the jurisdiction of Victoria City Council.

For aerial showmen (or their sponsors) who wished to draw a crowd, an important consideration was access by public transit. The B.C. Electric Company had actually created Oak Bay [Windsor] Park in 1890, and then owned it for many years; it had a track, and it was popular for rugby, field lacrosse, and pro baseball. This park and nearby Oak Bay beach were conveniently situated at the end of the Number 1 (Oak Bay) streetcar line. Willows Fairgrounds could be reached easily by the Number 11 (Willows) car.

The locations identified on this map will be more fully explained as each occurs in sequence. Except for Frank Sylvan (Chapter 3), the balloonists used site No. 1, near the modern Oak Bay Marina.

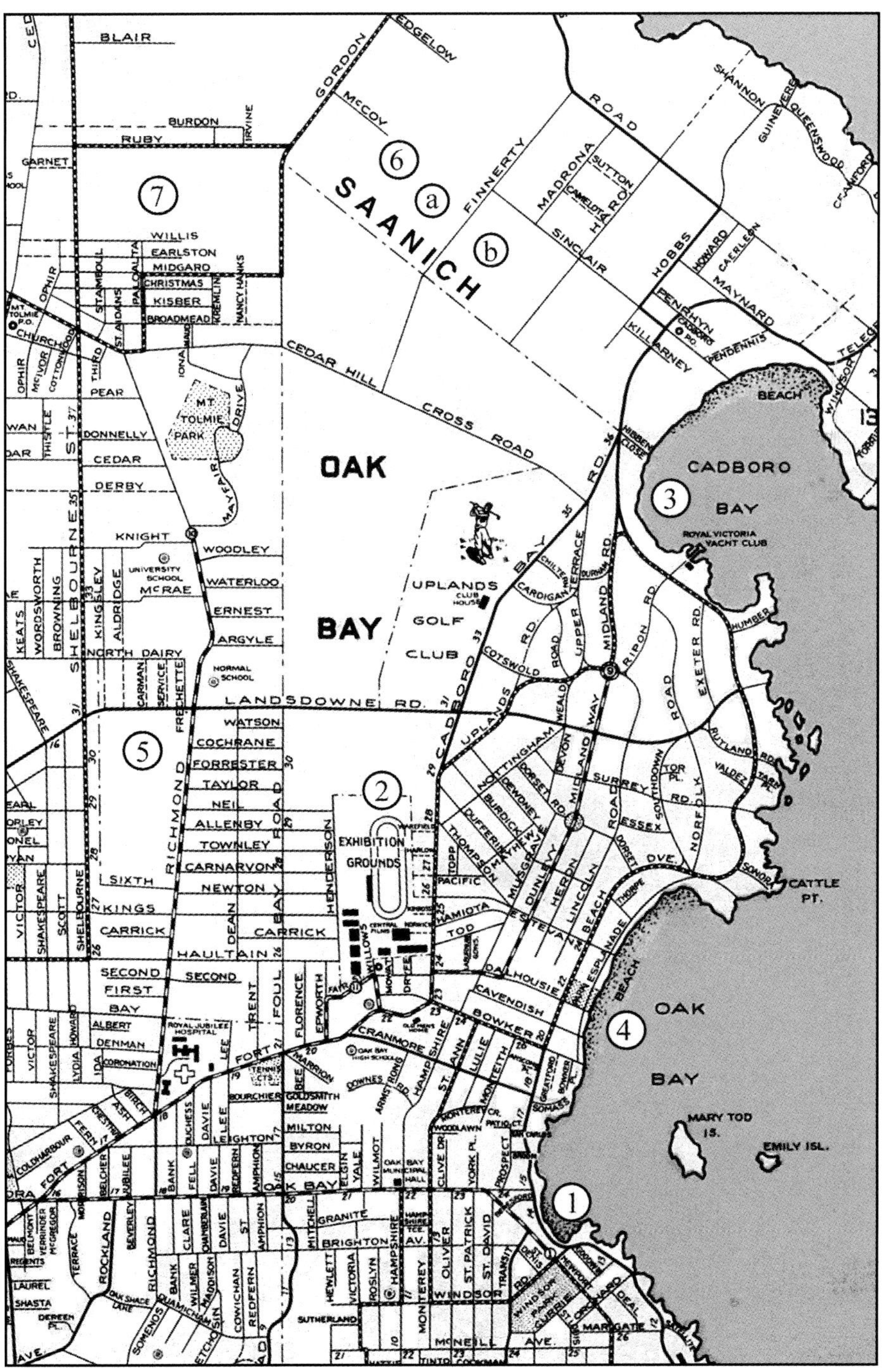


**LOCATIONS OF AERIAL ACTIVITY IN VICTORIA'S EASTERN SECTOR**

1. Oak Bay Waterfront and Oak Bay Park (now Windsor Park)
2. Willows Exhibition Grounds (in 1919, "Willows Aerodrome")
3. Cadboro Bay Beach (and Royal Victoria Yacht Club)
4. Willows Beach (used only occasionally by seaplanes)
5. Lansdowne Field (known earlier as "Deans Farm")
6. Gordon Head Airport (two stages: a.1930s, b.1940s)
7. Daffodil Strip (Joe Howroyd)

1938 Victoria & Island Publicity Bureau map (adapted). UVic Map Library G3514.54 V5 S6 1938 I8

## 3. Aeronautical Inventors and Barnstormers ▪ 1908–1913

### Frank Sylvan: Doctor, Aeronaut, and Would-be Inventor ▪ 1908

Acclaimed as the most successful to date of our city's fall fairs, the 48th Victoria Exhibition of 1908 drew to Willows Grounds a total attendance of about 50,000 over a five-day period, from Tuesday, September 22 through Saturday, September 26. The fair's traditional agricultural purpose was reflected in the judging of livestock, vegetables, fruit, and grain. A panel of outside experts had been imported from such metropolitan centres as Ottawa, Winnipeg, and Nanaimo. One judge commented that the quality of produce was the best he had ever seen—"much better than that shown at the Oregon State Fair." When grower T.A. Brydon trounced all his competition from the Lower Mainland, he spoke as the proud champion of Island agriculture: "Now will they tell us we can't cultivate good fruit here?"

Of course, there was also a wide variety of entertainment: horse races and parades, band concerts, vocal recitals, rifle shooting, and even Roman chariot races. The star attraction, however, was the daily balloon ascension and parachute drop performed by Professor Frank Sylvan. Each of these dramatic aerial events was reported breathlessly in the local papers. Here is the *Colonist*'s account of the first day's ascent on September 22:

> With the assistance of a number of the men, who, prompted by curiosity, crowded about the huge piece of canvas while it was being inflated, the preliminary arrangements were made expeditiously. Then the word was given and the aeronaut, clinging to a trapeze bar, shot far into the air, ascending with such rapidity that soon he appeared but a speck. He waved his arms jauntily at the people, who, craning their necks in the endeavor, watched his every movement. Soon he pulled the parachute drop. For just the fraction of a second he came towards earth with the velocity of a shot from a cannon. And then, very prettily, the canvas broke out in umbrella shape and he descended slowly and gracefully, alighting unharmed in full view of all. As there was absolutely no wind, the scene was spectacular in the extreme, and the professor reached terra firma within one hundred yards of the spot he had left but a few minutes before.
>
> —*Colonist* 1908.09.23: 2

As the week progressed, this accomplished showman introduced the far greater challenge of being completely shackled and—even more amazing—confined within a box:

> Before the balloon was set free, the operator was shackled, hand and foot, and securely nailed in a box, the latter having been made by D. Spencer Ltd. [*ed.* an impeccable local mark of integrity]

Historic postcard showing a Frank Sylvan balloon ascent in 1908.

City of Victoria Archives 98302-02-4335

He had ascended about five hundred feet before he managed to disengage himself, and it is estimated that he had gone four thousand feet when a small parachute was let drop with the box and the irons, which had bound him. Meanwhile the balloon had drifted directly over the grandstand. It was when in this position that the professor cut loose and began to sail down on his parachute. He alighted on the west side of the exhibition grounds, and a few minutes later appeared before the crowd and bowed his acknowledgement to the enthusiastic throng.

—*Colonist* 1908.09.25: 2

Because the few contemporary photographs show Professor Sylvan only a short distance above the ground, we should notice the astonishing height to which he soared on that occasion. After his Friday performance, Sylvan confirmed that, by his own conservative estimate, he had reached an altitude of five thousand feet before he started to descend:

At that height, he says, the exhibition buildings looked like tents, and people as large as flies. The appearance of the aeronaut and the balloon from the track was much the same. To the majority, it is safe to say, the professor did not look as big as the insect referred to and his conveyance hardly as large as an ordinary baseball.

It was at this juncture that the parachute was brought into service. "Five to one that the balloon hits the ground first," shouted an enterprising tout to the crowd. But there was none anxious to take up the bet even at such odds—the piece of canvas which was quickly losing its buoyancy was a strong favorite. The aeronaut floated down at an easy pace. He sailed directly over the grandstand and alighted free from harm three or four hundred yards away. Quickly gathering up his belongings he ran before the crowd and acknowledged the plaudits which were liberally tendered.

—*Colonist*, 1908.09.26: 26

During Sylvan's final performance on the Saturday, a gust of wind carried his balloon towards Willows Beach, and he appeared headed for an ocean landing. Whether by skill or good luck, he managed to reverse direction during his parachute fall, and proceeded to glide northward. The crowd last saw him as he disappeared behind some buildings on the slope of Mount Tolmie, where he must have managed to land without mishap.

From a *Colonist* feature article the following spring (see illustrations below), we learn—perhaps with surprise—that the intrepid Professor was a Victoria resident, who, when not soaring skyward, was actually a medical doctor. He was then hard at work, in March 1909, constructing a highly unconventional heavier-than-air machine, to be propelled by an engine slung beneath the frame and controlled by two horizontal parachutes. His inaugural flight was planned for Victoria, for, as he said, ". . . there's just as much air adapted for flying purposes around here as there ever was around Paris or the other haunts of noted aeronauts. And what's more, . . . it's better air."

He also liked the fact that the city was surrounded by so much water, since that element should prove less dangerous for unexpected landings.

Although Frank Sylvan was scheduled to fly this unlikely aircraft at Willows later in 1909, we hear no more about the project.

Yet another would-be aviator from Victoria, Mr. J. Watts, was reported by the *Colonist* in 1910 to be constructing a Bleriot monoplane in the city, having ordered an engine from England; but he too dropped out of sight.

Both men were eclipsed by a more successful local aircraft builder, William Wallace Gibson (next page). Still, the City of Victoria can be proud to claim Frank Sylvan as a resident, if only for his fearless balloon and parachute shows.

The photograph and drawing are reproduced from the *Victoria Colonist* (1909.03.23, Sunday supplement p. 3). The illustration of the aircraft may owe something to the artist's imagination, but probably had Sylvan's approval.

## William Wallace Gibson, Victoria's "Balgonie Birdman" ▪ 1906–1911
## Flight of First All-Canadian Aircraft: September 8, 1910

The story of this pioneer achievement begins in the early years of the twentieth century in the small prairie town of Balgonie, Saskatchewan, some fifteen miles east of Regina. There a young Scot named William Wallace Gibson—born in Ayreshire in 1874—had become preoccupied with flying, soon after reading in 1903 of the Wright Brothers' exploits at Kitty Hawk, North Carolina. Determined to build a heavier-than-air machine himself, he began experimenting—first with kites, and then with an actual flying model, which was powered by a reworked kitchen-blind spring roller. Meanwhile he was becoming a successful merchant, the owner of three general stores. Keeping his aviation experiments to himself for fear of ridicule, he became an affluent member of the community. Unwisely, however, he was persuaded to take on contracts to build a right of way for the Grand Trunk Railway, then under construction in his region. Instead of turning a profit, he lost $40,000, and was left almost penniless.

Photo courtesy David Maude

It was about this time that Gibson began to build a four-cylinder, four-cycle, air-cooled engine, which he took with him to Victoria, on deciding to move to the west coast in 1906.

Early in 1907, he met a miner named Locky Grant, a down-and-out prospector who had a promising-looking gold claim on Vancouver Island's west coast. No sooner had Gibson seen it than he gave Grant his boat, camera, field-glasses, rifle, and a hundred dollars, thereby becoming the proud owner of a gold mine. Procuring a small stamp mill and a water-wheel to run it, he hired Grant to help him produce a gold brick worth $1,200; this enabled him to sell the property for $10,000. Solvent once again, he made plans at once to resume construction of his first aeroplane.

Work began in 1908, in his James Bay home at 146 Clarence Street. He would encounter many difficulties. Perhaps the least of his annoyances was the reputation for eccentricity that he soon acquired locally. (Gibson himself reported that passers-by in the streets of Victoria would often flap their arms derisively as he walked past.) He formed a partnership with a Dave Hanbury, who was to contribute $5,000, but who was able to come up with only $500. Also, materials were hard to come by in Victoria at that time. The four-cylinder engine, he discovered, would be of little use. Because of its six-inch stroke and light construction, it would, to use Gibson's own words, "Jump like a chicken with its head cut off."

He set about to build a better engine, this time a six-cylinder air-cooled model. He went to Hutchinson Bros. Machine Works, and even though they had little faith in his design, they undertook to build it. In March 1910, therefore, another notable aviation first was achieved: the first successful aeroplane engine ever built in Canada.

It had a bore of 4.5 inches, with 4.5 inch stroke, and a battery-coil distributor ignition. The engine was unique in that two air screws were used: the front, direct drive turned a six-foot blade, while the rear, through a 2:1 gear system, swung a four-foot propeller. The object of this was to eliminate torque, which has continued to be a source of trouble to aircraft designers right up to the invention of the jet. This innovation was certainly another first. When tested, the engine ran beautifully, developing about 50 horsepower; it weighed 210 pounds.

The Gibson Twin-plane, as he called his aircraft, had an interesting design; a full-size replica now hangs in the British Columbia Aviation Museum at Victoria International Airport. It had two wings, one behind the other, each twenty feet from tip to tip and eight feet at their widest point. Its spruce framework was covered with blue waterproof silk, obtained from Jeune Brothers Tent and Awning Manufacturers in Victoria.

The wing design was basically what is known today as a "Gull Wing Type," and gave excellent stability. Practically everything on the plane was streamlined, and cambered to provide lift. Two streamlined ten-gallon gas tanks, gravity feed, were fitted above and to each side of the engine.

The "running gear," made from metal tubing attached to the main frame and using four bicycle wheels supplied by automobile dealer Tom Plimley, were likely the weakest design of the machine. The front elevator was eight by four feet, made of laminated cedar, and operated by a lever. Two rudders, also of laminated cedar, were operated by a shoulder yoke. There were no ailerons.

Early on the morning of Thursday, September 8, 1910, the momentous day arrived. In secrecy, with the aid of two helpers, Gibson moved his aircraft from its shop, by wagon, to the open field then known as Deans Farm. This is the same property off Lansdowne Road, between Richmond and Shelbourne, that would become Lansdowne airfield eighteen years later. Here the Twin-plane was erected, and a short but successful flight was made—probably just below the site of the modern Lansdowne Middle School. Unfortunately, the landing gear suffered some damage, necessitating repair.

By September 24, all was in readiness once again. At 4:00 a.m. the machine was trundled out to the same field, the motor warmed up; then Gibson signalled his helpers to let go. A slight cross-wind was blowing. Helped by this factor and by the downhill incline, the machine rose quickly after a run of only fifty feet. The cross-wind soon made itself felt in a negative way, as the aircraft drifted off towards some oak trees in the field. Not too familiar with the controls, Gibson leaned to the wrong side in an effort to overcome the drift; then, sensing an imminent collision, he shut off the motor and touched down, after being airborne for some two hundred feet. This was a remarkably long distance; as aviation historians have often noted, it was much farther than the Wright Brothers' first flight seven years earlier at Kitty Hawk. Unfortunately, the momentum of the landing carried the machine into a sturdy oak, with a bone-shaking crash. Severely bruised and shaken up, Gibson surveyed his badly damaged machine. Taking into consideration the approaching winter, he decided against further work at that time.

It should already be clear that Gibson was far more than just an inventive handyman: he was a skilled theoretician, who always kept abreast of current research. The winter months of 1910 gave him ample time to peruse the writings of Sir Hiram Maxims on the principles and theories of aeronautics. In the early spring of 1911, influenced by what he had read, he started work on a new craft called a Gibson Multi-plane. This craft had a series of narrow wings made entirely of spruce, with a more rigid frame. This new model included ailerons, as he now realized the need for lateral control. Because the two-propeller idea had not lived up to his expectations, he modified the engine to use a single eight-foot propeller with a six-foot pitch.

*British Columbia Historical Quarterly* 8.2 (April 1944) facing page 100

The Gibson Twin-plane: a replica hangs in the British Columbia Aviation Museum at Victoria International Airport. This was the aircraft that William Gibson flew in Victoria on September 8 and September 24, 1910. His later Multi-plane, built in Victoria during 1910–11, was flown successfully on a farm outside Calgary in the summer of 1911.

Gibson now parted company with Victoria. Wary of those treacherous oak trees, he received permission from Lieutenant-Governor Paterson to use his farm near Ladner, B.C., but adverse weather allowed only one short flight (*Colonist* 1911.06.02: 1). He then proceeded to set up his machine in Kamloops, B.C.; however, a fast-talking promoter persuaded him to continue to Calgary, where several flights were made. Apprehensive of the risks her husband was taking, Mrs. Gibson asked him not to fly any more. Accordingly, he accepted the help of a friend named Alex Jaff. Misfortune struck on August 12, 1911, when Jaff smashed up the machine while landing on a very rough piece of ground.

Virtually broke, having spent $20,000 on his experiments, Gibson moved to California, where he became a successful manufacturer of gold-mining equipment, most of which he designed himself. After his pioneer achievements had been researched and acclaimed in the 1940s by Frank Ellis, he was found to be still living in California, and he took his rightful place in the annals of Canadian aviation. He donated the first successful Canadian airplane engine to the National Aviation Museum in Ottawa (photograph, upper right).

This remarkably creative inventor died on November 26, 1965, at the age of 91. William W. Gibson's son represented him at a ceremony on September 8, 1985, when a 75th anniversary cairn was erected on the west side of Richmond Road, one block south of Shelbourne, to mark the site of his historic flight (see plaque, right).

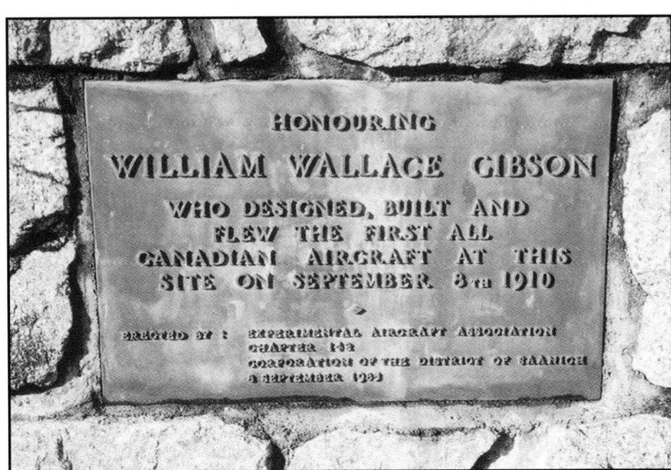


Honouring William Gibson's first flight on September 8, 1910, this cairn was erected by the Experimental Aircraft Association (Chapter 142) and the District of Saanich.

*The Balgonie Birdman* is the title of a charming puppet animation film about William W. Gibson, produced in 1991 by the National Film Board of Canada. The best account of his life and work is still Frank H. Ellis's article in *British Columbia Historical Quarterly* 8.2 (April 1944). Ellis gave Gibson a place of honour in his fine book, *Canada's Flying Heritage* (see Bibliography: Ellis 1961).

## J. Strobel and J.C. Mars: First Visiting Airship ▪ September 1909

The first airship to be seen in B.C. arrived in Victoria in September 1909. Owned by two Americans, J. Strobel and J.C. Mars, it was flown by the latter. A flight was planned at the Willows Exhibition Grounds for September 20, but it was aborted because of strong winds. On the next day the winds were blowing even more strongly, so J.C. Mars, good judgement overcoming the insults from somewhat hostile onlookers, cancelled the flight again.

By September 22, the weather had moderated and plans were made to ascend during the evening hours, when the wind had usually abated. A searchlight that was to be used developed troubles, and it was not until 10 p.m. that the airship made a circuit of the grounds. This was the first time that any Canadians had ever seen a powered aircraft flying over them at night.

The following day, after another delay that caused many of the spectators to leave, a second flight was made, and the next day (September 24) a third and final flight was made at the Exhibition Grounds. This time, over 12,000 spectators were on hand to witness the event—an astonishing number for a city of Victoria's size in 1909.

## Charles Francis Walsh: First Sustained Flight ▪ May 30–31, 1911

Postcard Collection of Donald M. Stewart

This souvenir postcard shows Charles Walsh in his Curtiss-Farman biplane at Willows on Tuesday, May 30, 1911.

When Charles F. Walsh of San Diego first flew at Willows Exhibition Grounds on May 30, 1911, everyone in the grandstand, reporters included, thought he was the advertised and eagerly awaited American aviator Fred J. Wiseman. After a brilliantly successful first flight, Walsh revealed that he was substituting for the better known Wiseman, and received a thunderous sustained ovation.

Walsh had arrived quietly with his entourage on a Monday, shipping his Curtiss-Farman biplane from Portland. On Tuesday, Willows was jam-packed, as Victoria's Fifth Regiment Band performed in fine style. That first flight lasted fifteen minutes, reaching a height of 600 feet:

> The first time he ascended after a run along the race track of about 200 yards, and then turned the machine into the centre field, at a part where the race track rails had been removed, and moving the front planes just as he reached the edge of the track commenced the upward flight. Making right across the ground above the horse and cow stalls at the southeast, he continued on over the Old Men's Home and round Cadboro Bay, returning to the front of the stand, where applause greeted his approach, and his answer was given by a wave of the hand. Walsh then circled again, and before returning took an excursion out over Smith's Hill, eventually turning and making a perfect landing at the point where he ascended.
>
> —*Times* 1911.05.31: 7

His second flight, at a much lower altitude, had a couple of new twists, inspiring an ominous prophecy:

> In dodging through the trees to the southwest of the Willows track, one of his planes [wings] struck the branch of a tree and cut it off with the cleanness of a razor. He brought the twig back with him and will keep it as a souvenir. In travelling past the grand stand he took an apple from his pocket and threw it at the judge's box, striking the corner, which illustrates that the aeroplane might be destructively useful when employed for military purposes.
>
> —*Colonist* 1911.05.31: 9

There were two more flights on Wednesday, May 31. Apart from expressing disappointment that Walsh was not able to take a passenger aloft, as advertised, the newspapers still sang his praises.

But when he decided that wind conditions would not allow him to fly at all on his third and final day in town, on Thursday, June 1—a special occasion aimed at children—Victoria's enchantment with aviation cooled considerably, and Charles Walsh was no longer viewed as the faultless and heroic pilot. His prudent caution about the wind was not appreciated, and he was sent on his way with grumpy press notices.

**Your Last Opportunity To See Birdman Walsh Toy With The Wind and Ride Upon The Clouds**

TODAY IS CHILDREN'S DAY

ONLY "TWO BITS" FOR YOUNGSTERS AND "FOUR BITS" FOR ADULTS TO WITNESS AVIATOR'S LAST PERFORMANCE HERE IN HIS HEAVIER THAN AIR CURTISS-FARMAN BIPLANE AT EXHIBITION GROUNDS THIS AFTERNOON.

An attractive programme prepared.
One of those present will be invited to take a trip through the air.

The first flight will take place at 2:30 o'clock sharp. Birdman will fly, no matter what the weather conditions. He likes the wind.

**REMEMBER**

**50c General Admission Children 25c**

And it's the last invitation you'll have.

*Colonist* 1911.06.01: 9

## The Disappointing Visit of J.V. Cavanaugh ▪ March 16, 1912

If Victoria had seemed a little uncharitable to the admirably successful aviator Charles Walsh in 1911, his treatment was gentle when compared to the reception given J.V. Cavanaugh the following year. Admittedly, Cavanaugh's visit was a flop.

**Spectacular Entertainment**
**Aviator J. V. CAVANAUGH**
The Famous Bird Man, Gives An
**AVIATION EXHIBITION AT**
**VICTORIA FAIR GROUNDS**
This Afternoon at 3:30 o'clock.
**Record flight is assured. This is a sight of a lifetime and you should be a witness.**
**Admission 50¢. Children under 10 years free.**

*Colonist 1912 03 16: 9*

Walsh had rashly advertised that he would fly "no matter what the weather conditions." Cavanaugh went one step further: "Record flight is assured" (see advertisment). The event was scheduled for the Willows on Saturday, March 16, 1912.

A brief notice buried on page 16 of the next morning's *Colonist* carries a starkly eloquent headline: AEROPLANE STRIKES FENCE AND FALLS. A large crowd had been attracted to the Exhibition Grounds on a cold and blustery day, "expecting to witness aviator Cavanaugh play some of the pranks of the seagull. . . . It is mild criticism, indeed, to say that few of those present went away satisfied."

In photographs we see the American barnstormer's handlers pushing his aircraft across the infield almost as if coaxing it to take flight. On takeoff, the machine failed to clear the surrounding fence, smashing the stabilizer from the front of the plane; the pilot luckily escaped injury. Requests for the refund of admission fees were not favourably received.

*Peter Corley-Smith Collection*

## The Pride of British Columbia: William Stark ▪ May 24–25, 1912

For Victoria's May 24th Celebration in 1912, the headline attraction was Vancouver's Billy Stark, who had just recently become a British Columbia celebrity. On April 13, he gave a solo flying demonstration at Minoru Park in Richmond, and on April 24 he flew with his wife Olive, who thus became the first woman aeroplane passenger in Canada. Trained in San Diego, Stark was a pioneer Canadian barnstormer who could bear comparison with any of his American counterparts. Our summary will be brief, since his story has been well told by Frank Ellis (1961) and Peter Corley-Smith (1989).

On his Victoria visit, Billy took off from Oak Bay Park, now Windsor Park, and flew his Curtiss pusher biplane for some twenty minutes, including a pass over Beacon Hill Park (see photograph below). On his landing back in Oak Bay, the crude braking device on the front wheel would not hold, and the machine collided with a fence. Because the damage was minor, repairs could be made for a second flight the following day. This time he soared into the air for a fifteen-minute flight, followed by a safe landing.

Left: May 24, 1912, Victoria Day. William M. Stark flies over Beacon Hill Park during a two-day visit to Victoria.

Below: July 1, 1912, Dominion Day. Billy Stark attracts a huge crowd on a visit to Armstrong, giving the first exhibition flight in the B.C. Interior.

## Alys McKey Bryant and John Milton Bryant ▪ August 5–6, 1913

In these years just before World War I, most Canadian cities of reasonable size were enjoying exhibition flying on a fairly regular basis, with the great majority of performances given by travelling aviators from the United States. John and Alys Bryant, a husband-and-wife team from California, came to Victoria after a notable exhibition at Minoru Park on Vancouver's Lulu Island, now part of Richmond. When both husband and wife had flown their Curtiss pusher biplane there on July 31, 1913, Alys earned the distinction of recording the first flight in Canada by a woman pilot.

Their machine was shipped to Victoria, where Alys was planning to make a similar pioneer flight from Willows Race Track on August 5. It was a day when the winds were gusting off the ocean, and she had great difficulty in keeping the craft steady. Instead of circling over the city as she had planned, she was forced to turn back towards her starting point. As she said afterwards, this landing at Willows ended one of the worst flights she had ever made. But there was far worse to come.

Although the winds had not abated on the following day, Johnny Bryant was still determined to be the first man to fly over any city in British Columbia. By adding a single float in lieu of wheels, he converted the machine into a seaplane, and took off from Cadboro Bay.

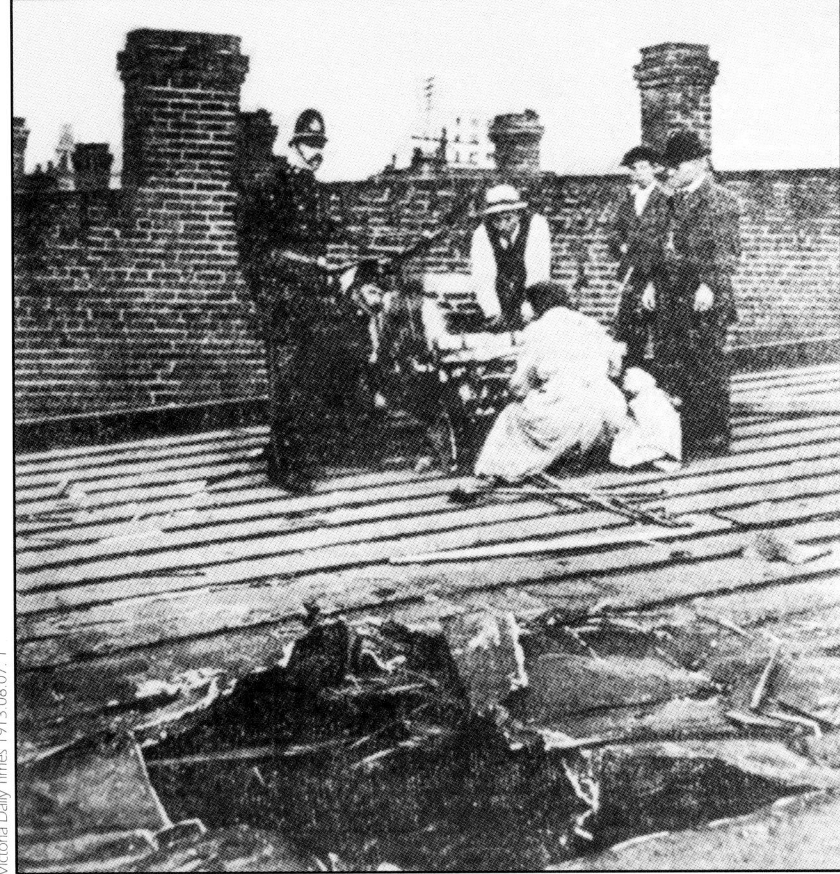

Victoria Daily Times 1913.08.07: 1

After testing the machine's handling capabilities with a short flight, he took off a second time and headed directly for downtown Victoria, where assembled crowds craned their necks, looking skyward to gaze in wonder at the sight. Crossing over the city, he landed in the Inner Harbour, where he was given a rousing welcome by spectators lining the wharves. Despite severe gusts of wind, he took off again shortly before 6 p.m. After circling for height, he flew back over the downtown business district at an altitude of about 800 feet. When directly over City Hall, his machine plunged suddenly and sharply. Although Bryant could be seen to struggle valiantly for control, his efforts were to no avail. About 200 feet above the ground, the wing broke away, and moments later his machine crashed through the roof of the Lee Dye building in Chinatown, between Fisgard and Cormorant at Theatre Alley. Unconscious and barely alive, Johnny Bryant soon died of multiple fractures, becoming Canada's first aviation fatality.

## PART TWO

# THE GREAT WAR AND THE POSTWAR DECADE

## 1914–1929

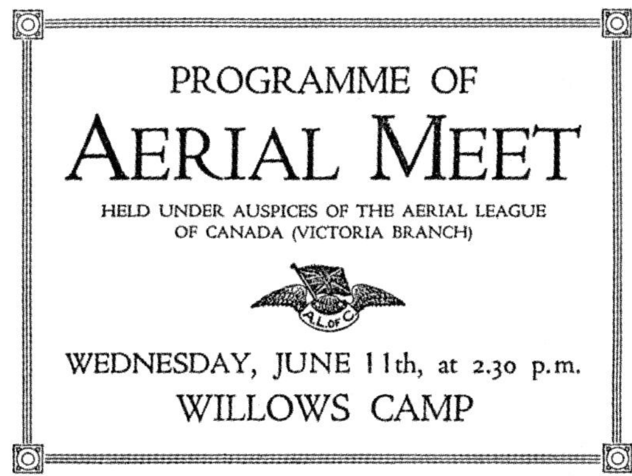

PROGRAMME OF

AERIAL MEET

HELD UNDER AUSPICES OF THE AERIAL LEAGUE OF CANADA (VICTORIA BRANCH)

WEDNESDAY, JUNE 11th, at 2.30 p.m.

WILLOWS CAMP

# 4. World War I ("The Great War") ▪ 1914–1918

It would be straying far beyond our scope to explore the aerial history of World War I. There have been many books on Canada's conspicuous role in the skies over Europe; a fine recent study is David L. Bashow's *Knights of the Air: Canadian Fighter Pilots in the First World War* (2000). What is important to stress here is that those distant events in Europe had a profound effect on B.C.'s provincial capital, touching almost every family on southern Vancouver Island. Victoria was a city that considered itself a proud and loyal outpost of the Empire, and young men enlisted in vast numbers. In a school that then had an annual average of fewer than 300 male students, Victoria High alone sent over 450 alumni volunteers to the Great War—and this was only a small minority of the local contingent. Women draped in the Union Jack walked beside files of uniformed recruits who marched down Government Street to the CPR Wharf. The *Princess Victoria* and other coast steamships were crammed with load after load of young volunteers heading off to combat, many hundreds destined never to return.

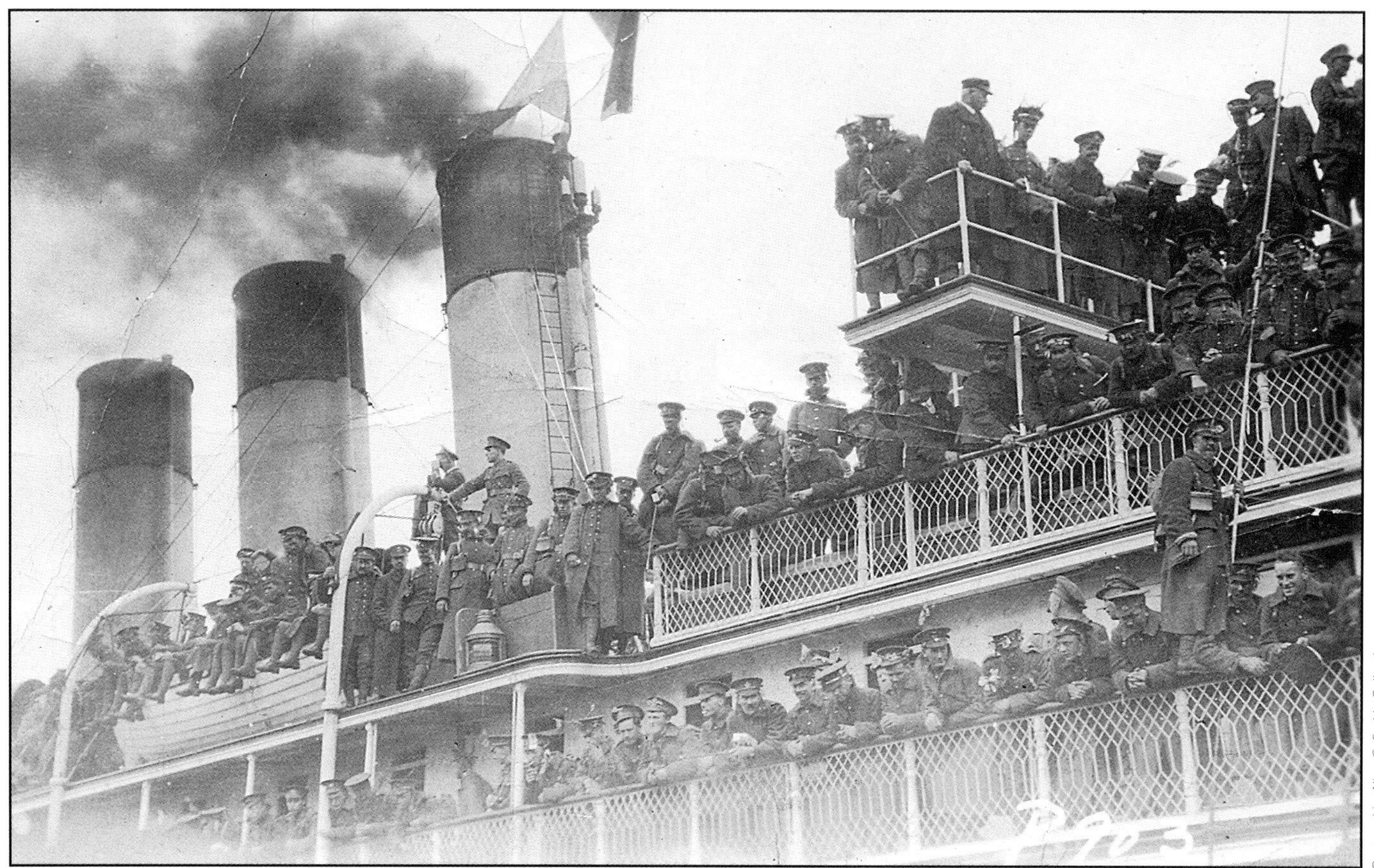

Postcard in Alice C. Smith Collection

Departing Canadian troops wave goodbye as the *Princess Victoria* leaves the CPR Wharf in the Inner Harbour.

What began as a grand patriotic adventure became a grim and bloody conflict. Victoria High School would count 86 former students among the dead. The majority of Canadian casualties were suffered on the ground, at battles whose evocative names still arouse mingled emotions of pain and pride—Ypres, Passchendaele, Cambrai, Vimy Ridge.

But Canada also played a disproportionately large role in the aerial campaigns of the Great War. Historians have calculated that 22,812 Canadians served overseas with British flying forces—the Royal Flying Corps and the Royal Naval Air Service, which merged on April 1, 1918 to form the Royal Air Force; of this total figure, 13,160 served as aircrew (Bashow 2000, p. xi).

One can only guess at the number of World War I aviators who came from Victoria, but there was clearly a large contingent. Of those 86 Victoria High students who gave their lives, at least ten were in one of the flying services. We know the names of many who survived and maintained their interest in flying after they returned home, thus influencing the course of civil aviation in Canada over the next two decades.

There has seldom been a modern war that did not result in a surge of technical progress. In this respect, the First World War may be viewed as the conflict in which aviation came of age—if that is not too glorified a way to describe the practice of shooting other human beings out of the sky. The primitive flying machine of the pre-war decade had little in common with the manoeuvrable fighter aircraft of 1917 and 1918. The pace of technical achievement did not go unnoticed. People around the world became fascinated with aviation, their interest fired by the highly dramatic style of close combat in the skies over the Western Front. Those aerial dogfights were romanticized as grand theatre, pitting courageous Allied flying aces against their brilliant and not unworthy German adversaries. Despite the appalling waste of human lives, the skill and heroic bravery of these young aviators very much deserves our admiration and respect.

At the time, Canadians took particular pride in being able to claim the two top-ranking British aces, Owen Sound's Billy Bishop and Nanaimo's Raymond Collishaw. British Columbia had a legitimate claim on another highly decorated fighter pilot, Ottawa native Donald MacLaren, who would play a variety of important roles in post-war aviation.

Billy Bishop's name is still familiar to most Canadians, but only aviation enthusiasts are now likely to know anything about Collishaw or MacLaren. Therefore we shall summarize very briefly the wartime achievements of these two British Columbia aces. Then we shall look at the poignant stories of Robin Gray and Reginald Litchfield, two native Victorians whose young lives were rich with promise, but who shared the tragic destiny of so many courageous Great War volunteers.

Above: Lt.-Col. Raymond Collishaw
Below: Maj. Donald R. MacLaren

Photos: Drew (1930)

Lieutenant-Colonel Raymond Collishaw, C.B.E., D.S.O. and bar, D.S.C., D.F.C., Croix de Guerre, came out of World War I with 60 aerial victories, second only to Bishop's incredible 72. What set Collishaw apart from most other Canadian aces is that he flew for the Royal Naval Air Service; because that force shunned the limelight, this Nanaimo-born pilot was virtually unknown to the world until the war was over.

Collishaw's life was one continuous adventure. In pre-war days, he had been second officer on a ship that ran between Victoria and Skagway. Then he joined and survived Scott's Antarctic Expedition. After the 1918 armistice, he continued to fly war missions with the R.A.F., first in Russia and then throughout the Middle East. He never returned to civilian life in Canada; as an Air Commodore in World War II, he served with distinction as commander of the so-called Desert Air Force.

Like Raymond Collishaw, Donald MacLaren was born in 1893. His family moved from Ottawa to Calgary in 1899, and then to Vancouver in 1912. Enlisting in 1916, he received all his R.F.C. training at Ontario's Camp Borden. Because he will be mentioned often in later chapters, we shall here give only a bare summary of his war record.

Major Donald R. MacLaren, D.S.O., M.C. and bar, D.F.C., Chevalier of the Legion of Honour, Croix de Guerre, was credited with 48 aerial victories and 6 balloons destroyed, all during a dazzling six-month period in 1918. He finished the war in sixth place among British air aces, having flown in combat for a much shorter time than those who surpassed his record.

Both Raymond Collishaw and Donald MacLaren were early inductees into Canada's Aviation Hall of Fame.

George Robert Gray was known always as Robin. "Roslyn," his family home in Victoria West, still overlooks the Gorge at 1135 Catherine Street—one of the City's finest heritage houses. His father Andrew Gray arrived in Canada from Scotland in 1871, at age 19; he became a highly successful marine engineer, part-owner of the Victoria Machinery Depot, manager of Albion Iron Works, and founder-proprietor of Marine Iron Works on Pembroke Street.

Courtesy of Rear Admiral William A. Hughes

Lieutenant George Robert (Robin) Gray. Born Victoria, May 7, 1898. Died when shot down over France, October 31, 1917.

Robin's two older brothers, Jack and James, served overseas; Jimmy Gray was an accomplished pilot, and will make a major appearance later in this section. Andrew and Mary Gray also had a daughter, Elizabeth.

Robin was a very good student, a champion swimmer, and by all accounts an outstanding young man. He was 18 when he graduated from Victoria High School in June 1916, and only 19 when he was killed in action as a pilot in the Royal Flying Corps—shot down over France on October 31, 1917. His mother, Mary Gray, who lived until 1953, never got over the loss of her youngest son.

Robin Gray had been recruited by the RNAS into the Royal Flying Corps. After about 20 hours of training in the U.K., Lieutenant Gray was posted to No. 84 Squadron in France on October 6, 1917. On his third patrol, while flying an SE5, he was shot down on October 31, 1917; he landed his plane, but died of his wounds shortly afterwards.

Lieutenant G.R. Gray is buried in Tourcoing (Pont-Neuville) Communal Cemetery, and his name is also inscribed on his parents' tombstone in Victoria's Ross Bay Cemetery.

*(Our thanks to Rear Admiral William A. Hughes and Dr. John A. Gray.)*

---

On a wall of the Victoria High School Archives, there is a framed photograph of Reginald Litchfield, conspicuously placed among a set of historic pictures that commemorate former students who gave their lives in World War I. When present-day students visit the room to look at the artifacts and mementos of a bygone era, they almost always comment on what a strikingly handsome young man he appears to have been.

Victoria High School Archives

Cadet Richard William Reginald Litchfield. Born Victoria, July 9, 1895. Died in a training accident, May 2, 1918.

When it is pointed out that Cadet Litchfield was killed in a training accident during the last year of the war, before he had even qualified as a pilot to be sent overseas, visitors sometimes express the opinion that this was an even more cruel loss than a death in action. But of course training accidents have been and perhaps always will be regrettably inevitable in the course of preparing student pilots; later in this book we shall note the horrific number of young RAF trainees killed at Patricia Bay Airport during World War II.

The airplane accident that killed Reginald Litchfield occurred at the School of Aerial Gunnery, Beamsville, Ontario. A senior flight instructor at Beamsville was Jack Clemence, whom we shall see as a post-war pilot with the Aerial League of Canada (Victoria Branch).

Cadet Litchfield's parents were Henry Oswald Litchfield, a bookkeeper, and his wife Eliza (née Daniels), who lived at 1147 Hilda Street in Victoria's Fairfield District. They had been married in Victoria on October 2, 1894, and Reginald was born the next year. Their son is buried in Ross Bay Cemetery.

## 5. The Aerial League of Canada, Victoria Branch ▪ 1919–1920

Within months of the Armistice on November 11, 1918, both Vancouver and Victoria witnessed a flurry of aeronautic activity. The main impetus, naturally enough, was provided by the large group of former wartime pilots who flocked to the west coast. A few of these were new immigrants from Britain, or transplants from other parts of Canada. The great majority, however, were homecoming British Columbians, including a surprising number of native Victorians. Many appeared eager to help Canada develop new opportunities in civil aviation, even if that required some prodding.

By no means all these Royal Flying Corps and RAF veterans wanted careers in aviation. Some had good positions waiting for them in Victoria, like Robert H.B. Ker, scion of the Brackman-Ker Milling Company; or Herbert Sandham Graves, a future editor of the *Colonist*, who was already well launched on a career in journalism; or Emerson L. Smith (*see below*), a well-liked Victorian who would succeed his father as head of Ship Chandlers (McQuade's) Ltd.

Other more obscure young veterans were badly in need of a job, and hoped to parlay their well-honed flying skills into profitable employment. This group included soon-to-be acclaimed aerial heroes like Robert L. Rideout and W. Harry Brown. The slow pace of development in Canada would cause half a dozen of these accomplished pilots to head south eventually for better opportunities in the aviation-conscious west coast of the United States.

Meanwhile, a Canadian flying boom was fostered by the availability and low cost of reliable aircraft. Almost without exception, the machine of choice was the versatile Curtiss JN-4 Canuck, known affectionately (like other Curtiss JN models) as the "Jenny." Over 1,200 of these biplanes were manufactured during the war by Canadian Aeroplanes Ltd. of Toronto, and had proven highly successful, in both Canada and the U.S.A., in training recruits for the Royal Flying Corps. With the surplus stock of this aircraft now a glut on the market, a new or almost new Canuck could be bought for as little as $2,000.

Almost no veteran could ever have managed such a purchase on his own. Fortunately, the Government helped create a national organization—the Aerial League of Canada—that enabled eager ex-military pilots to band together to form local branches, collectively buying these aircraft so as to polish and exploit their flying skills.

Photos courtesy of Austin Smith

One returning Victoria pilot who did not choose to extend his flying career was Lieutenant Emerson L. Smith, who had earned official status as a British ace by downing at least seven German aircraft. Shot down himself behind German lines after an aerial dogfight on October 17, 1917, he was accosted in a field dressing station by the mighty Baron Manfred von Richthofen, who angrily rebuked him in English for having just shot down his ace pilot Hermann Goering, future Luftwaffe head (*Colonist Islander* 1959.09.20: 2–3). Here we see Emerson Smith (L–R) as recent graduate of RFC training, as prisoner of war in 1917–18, and as carefree owner-manager of Ship Chandlers (McQuade's) Ltd., Victoria. Mr. Smith looks pleased not to have pursued a career in aviation.

The Aerial League had already been active in Montreal and Toronto; but the two branches created in Victoria and Vancouver were the first in western Canada. Soon these branches encouraged others to form in Alberta and Manitoba. The League's first and most important goal was "to demonstrate the dependability of flying, viewed from a commercial standpoint" (*Colonist* 1919.04.06: 17). This statement expresses the League's deep faith in the future of commercial aviation, a conviction not shared in early 1919 by all members of the population.

However, there is no sign that even sceptical Victorians disapproved of the fledgling Aerial League. With the help of frequent and positive newspaper support, the Victoria Branch took just two months (March and April 1919) to make its dream a reality. The business community was warmly sympathetic, though there was no immediate offer of financial help. Premier T.D. Pattullo encouraged a League suggestion that the Forest Branch might develop a program of fire detection from the air (*Times* 1919.04.12: 7).

When League members bought their first aircraft with their own money, local citizens applauded and eagerly awaited the plane's arrival. Victoria City Council gave cautious approval for "Willows Camp"—its wartime label—to become an aerodrome, with part of the Horse Show Building to be remodelled as a hangar.

Only one alderman, it seems, chose to look on the gloomy side:

> These aeroplanes are a new thing for Victoria, and are apt to run amuck occasionally. They are at such times very destructive. If one of them was to kill half a dozen people or more we might have to pay.
>
> —*Times* 1919.04.23: 13

Because the spring of 1919 was so crammed with newsworthy events, we shall summarize the more interesting and important of these in diary form.

**MONDAY, MARCH 10, 1919**

The Victoria Branch of the Aerial League of Canada (ALC) is officially formed, chooses its first directors, and elects its officers:

Lieut. W. Harry Brown
Lieut. Gordon A. Cameron
Lieut. L. Louis Grant, *Vice-Pres.*
Capt. James Gray
Lieut. Harry D. McDiarmid, *Secretary-Treasurer*
Lieut. Robert L. Rideout
Mr. P. Thompsett, *President*

Not mentioned initially is Lieut. H. Sandham Graves, who was added soon as a director, with a special responsibility for publicity. On April 13, Graves anchored a full-page aviation feature in the *Colonist*.

**THURSDAY, APRIL 3**

Lieut. Harry McDiarmid delivers a stirring and highly perceptive address to the Rotary Club:

> After supplying 20,000 aviators, 60 per cent of the personnel of the British air force, it would be deplorable if Canada should lose the great asset which those men constitute simply by neglecting its opportunity now to find new channels for their usefulness. Canada has not yet paid a red cent in the cause of aviation, yet every other country is planning peacetime development in aerial work. If Canada does not show that she means business in the air, it will soon be too late to start.
>
> —*Colonist* 1919.04.04: 7

**MONDAY, APRIL 28**

The Victoria ALC's first aircraft is towed from the steamer *Charmer* through the streets of the city, to be assembled and rigged at the Willows Horse Show Building under the expert supervision of Jack Clemence. It is hoped that the new Curtiss JN-4 will be ready to fly by Saturday.

Gordon Arthur Cameron was President of his 1915 matriculation class and a fine athlete at Victoria High School.
Victoria High School Archives

In the evening, 600 guests attend an Aerial League gala Dance-Cabaret in the ballroom of the Empress Hotel.

**TUESDAY, APRIL 29**

A massive civic luncheon is held to open the Victoria and Island Development Association's fund drive, a community effort that the Aerial League has agreed to support as a goodwill gesture. In a brief address, Lieut. Bob Rideout amuses the gathering by poking fun at City Council's view that an aircraft is apt to run amuck: "He calmed their fears by saying that the League would keep the plane chained up and muzzled" (*Colonist* 1919.04.30: 5).

**FRIDAY, MAY 2**

7:02–7:22 p.m. An evening test flight of the League's new two-seater Jenny is carried out by Pilots Clemence and Rideout. "It climbed to an altitude of 3,500 feet and in the progress of the flight a number of turns, vertical banks, stalls and wing-tipped 'revs' were made. The two pilots who alternately flew the plane put it through a very rigorous test, and when they landed at 7:22 o'clock reported everything satisfactory" (*Times* 1919.05.03: 4).

# The Aerial League of Canada, Victoria Branch

Gordon A. Cameron Collection

Officers and Directors of the Victoria Branch, Aerial League of Canada, pose in front of their Curtiss JN-4 biplane, recently named *Pathfinder*. The photograph was taken at the Willows Aerodrome, probably on May 7, 1919.

Even a grainy newspaper image can bring an unfamiliar name alive. Sussex-born Percy Thompsett, who emigrated to Victoria in 1912, was the oldest of this group at 35. Louis Grant was born on August 8, 1895 in Union, B.C. Harry McDiarmid had been a wartime Observer, like Gordon Cameron; born in 1897 in Lindsay, Ontario, he was a Victoria student—classmate of Gordon Cameron's—prior to his enlistment in April 1916.

Percy Thompsett, President

L. Louis Grant, Vice-President

Harry D. McDiarmid, Secretary-Treasurer

Senior Pilot Jack Clemence was an aerial virtuoso—the stunt-flying star of every air show. An older man (born in Uxbridge, Ontario, June 12, 1888), he had spent the war years as an instructor in Canada.

John A. Clemence

Robert L. Rideout

W. Harry Brown

James Gray

To the right are three other ex-Great-War pilots in the Victoria Branch. Bobbie Rideout was born in Newfoundland in 1897, but came to Victoria as a boy of 10. Harry Brown (b.1894) and Jimmy Gray (b. 1891) were both native Victorians who attended Victoria High School. Lieutenant William Henry Brown, M.C., had official status as an Allied ace, having been credited with nine aerial victories on the Western Front.

Newsclipping photos from Gordon A. Cameron Collection

**SATURDAY, MAY 3, 1919**

The Victoria Branch's new aeroplane makes its first flight over the city with Jack Clemence as pilot and Sandham Graves as passenger. In the process, the plane drops 5,000 leaflets to advertise the Victoria and Island Development Association—"the first occasion in the history of the West that this form of advertising has been indulged in." (This may seem an unlikely claim, until we realize that "West" means western Canada, not Western Civilization.) Then the plane flies low over Royal Athletic Park, so that a bag of eight "suitably inscribed" souvenir baseballs may be dropped onto the field, to be auctioned off to the highest bidders (*Times* 1919.05.03: 1). Shrewd businessmen!

Born in Ireland, Herbert Sandham Graves was educated at Victoria High School and Vic College before the war. By 1919 he was well along a career path that saw him become Editor of the *Victoria Daily Colonist*, when this picture was taken. Lieut. Graves wrote a book about his wartime flying experience: *Lost Diary* (Victoria: King's Printer, 1941).

Photo courtesy the *Times Colonist*

**TUESDAY, MAY 6, 1919**

The Victoria Branch's Curtiss JN-4 is flown shortly before 7 o'clock this evening by pilot Bob Rideout, with YMCA physical instructor J. Thompson as his passenger. After the plane descends to a height of 600 feet, Thompson drops two baseballs into Athletic Park, and one of these is used in the game (*Colonist* 1919.05.07: 6).

The aircraft will be named *Pathfinder* on May 7.

**WEDNESDAY, MAY 7, 1919**

On Tuesday, the *Times* had run a banner advertisement (left), promising "flying, stunting, and passenger carrying." The Willows Aviation Meet of May 7, 1919 was an historic first for post-war British Columbia. Even so, the *Colonist* may be stretching our credibility to describe it as "the first event of its kind ever held in Canada." While heaping praise on the magnificent flying demonstration, the writer fumes about the stinginess of Victorians who did not pay the modest 50¢ admission fee.

Gordon A. Cameron Collection

Lieut. Harry McDiarmid (left) and Lieut. Robert Rideout in *Pathfinder*, May 1919. This photo may have been taken at the Aviation Meet on May 7, when the pair gave a bombing exhibition (see previous page).

On Sunday, May 11, the *Colonist* ran a full-page spread, with other pictures taken at this Willows show.

The photo below has an album caption "Air Meet, Willows 1919." It is not more precisely dated, but the aircraft appears to be the original *Pathfinder*. If so, it must be May 7.

Gordon A. Cameron Collection

The *Colonist* report, probably the work of Lieutenant Graves, appeared the following morning:

> While the exhibition of flying was excellent even to the trained eye of the military aviator, members of the League were genuinely disheartened by the apparently lukewarm support which the citizens gave. Many parties in expensive motor cars parked their cars on streets adjacent to the Willows grounds or toured along the highways that commanded a favorable view of the performance, but took care not to enter the gate and contribute to the meagre funds of the League, which is trying to make aviation "go" in Victoria against tremendous odds. The result was that from a financial standpoint the meet was not a success. The gate receipts totalled $325.
>
> —*Colonist* 1919.05.08: 7

The writer describes a variety of ceremonial events:

> The christening of the Curtiss plane by Miss Gwendolyn Richards, whose suggestion that the machine be called Pathfinder found general commendation, was the first event of yesterday's programme. A bottle of champagne swung from the propeller was smashed with a hammer by Miss Richards, thus emulating in a fashion the age-tried tradition of ship launchings. Miss Richards was presented with a huge box of chocolates as a gift from the Aerial League. [Miss Richards was a teenager.]

Major-General R.G.E. Leckie, C.M.G., gave the ceremonial first crank of the propeller, and the show was under way.

The *tour de force* performance by Jack Clemence is described at right. The show also included a bombing exhibition by pilot Bobbie Rideout and observer Harry McDiarmid: ten bags of flour were dropped, and they burst satisfactorily. (The *Colonist* account added, "The automobiles standing outside the park were the particular objective.") The afternoon ended with a short series of complimentary passenger flights. Except for the lack of revenue, the meet could scarcely have gone better.

## FRIDAY, MAY 9, 1919

"PATHFINDER" LOOPED IN FORTY-MILE GALE: AEROPLANE DEMONSTRATED THE PRACTICABILITY OF FLYING IN ADVERSE WEATHER

—*Times* headline, Saturday, May 10

*Pathfinder* performs various stunts in a forty-knot gale. Jack Clemence shrewdly exploits the storm to present a vivid, dramatic demonstration of aerial stability. Clearly a very skilful pilot, he makes his point in a half-hour show over Athletic Park, with a loop, stall turn, spinning nose dive, and a roll.

### John A. Clemence, Aerial Virtuoso

In reporting the Willows Meet of May 7, the *Victoria Times* provides a precise and expert catalogue of Jack Clemence's stunt repertoire:

> After a vertical tank and a few turns, the nose of the plane went down and a perfect loop-the-loop was made. This was the first time this stunt had been performed west of Toronto. Immediately afterwards, the "Pathfinder" did a half roll and then two complete rolls, followed by another loop. . . . [Suddenly,] a spinning nose dive, which caused shrieks amongst the ladies . . . . another loop, a cart-wheel turn, the left and right Emmelmann turns, another spinning nose, a splendid side-slip and to complete a most sensational flight, Clemence made a three-point-three landing, which means A1.
>
> —*Times* 1919.05.08: 15

That same day, the *Colonist* reveals that our pilot "is recognized as one of the foremost aviators," and that he was kept at Beamsville training station during the war because of "his special prowess in handling aircraft" (*Colonist* 1919.05.08: 7):

> In 'stunt' flying Clemence is said to have few, if any, superiors in the Dominion, and those who saw him yesterday saw the best.

> When the 'Pathfinder' was stunting over the ball park the crowd lost interest in the game, and the players got the aeroplane neck.

The stunt intended for Friday was postponed:

> The passenger will drop a baseball, which Ted Easterly, the Tyees' backstop, will attempt to catch, provided it falls within the park.

Gordon A. Cameron Collection

With propeller spinning, *Pathfinder* is almost ready for takeoff.

**SATURDAY, MAY 10, 1919**

VICTORIA'S BEAUTY MULTIPLIES WHEN SEEN FROM COCKPIT OF "PATHFINDER": first-hand account of an aerial tour in *Pathfinder* by Archie Wills.

—*Times* 1919.05.12: 7

This is an entertaining account of a Victoria sight-seeing tour (his first flight), for which *Times* reporter and artillery veteran Wills was required to wear "a thick helmet, a pair of goggles and a heavy flying coat" (see photo at right).

Archie Wills will appear again in our narrative. Years later, Archie recalled dropping two baseballs into Athletic Park that day—and missing the Park on both attempts!

Gordon A. Cameron Collection

May 10, 1919: a rare picture of pilot Jack Clemence (centre left), about to give *Times* reporter Archie Wills (centre right) an aerial tour of Victoria. They are flanked by the mechanics who recently helped Clemence assemble *Pathfinder*.

**MONDAY, MAY 12, 1919**

Two Vancouver Aerial League men fly from Vancouver to Victoria and back. Capt. Alfred Eckley achieves a huge "first": a round-trip flight between Vancouver Island and the mainland. His passenger was a mystery man—perhaps Ernest Hall, but Victoria newspapers identify him as Carl Dull, and the *Vancouver Province*, C. All.

> The Vancouver plane arrived unexpectedly yesterday afternoon at 5.45 o'clock after a trip of one hour and twenty-two minutes. When the machine was seen above Victoria it was believed to be the 'Pathfinder' out for a spin. Thus to the Vancouver airmen belongs the credit of being the first to negotiate the air passage from the Mainland to Victoria.
>
> —*Times* 1919.05.13: 7

The Vancouver aviators delivered a diplomatic letter from Vancouver Mayor Gale to Victoria Mayor Porter, and messages to the managers of the *Times* and *Colonist*. The two men were received with courtesy, and given dinner at the Empress Hotel. Leaving the Willows aerodrome at 7:12 p.m.—quick dinner!—they landed at Minoru Park at 8:06, exactly 54 minutes later, establishing the benchmark record for the Georgia Strait crossing.

No mention is made of the fact that the Vancouver Branch had brilliantly but rather meanly scooped its rivals in Victoria. It was a poorly guarded secret that the Victoria Branch planned to make exactly the same flight, in the opposite direction—the very next day! If it really was former Victorian Ernest Hall who flew with Eckley, he had some cause to conceal his identity.

**TUESDAY, MAY 13, 1919**

Victoria pilots strike back in *Pathfinder*, claiming at least four new records. Lieut. Jack Clemence and Lieut. Sandham Graves fly from Willows Aerodrome to Minoru Park and back. The one-way record is reduced to 42 minutes; the round trip flying time, to 158 minutes. They also set records for Vancouver to New Westminster and New Westminster to Victoria. The validity of these records was immediately challenged in Vancouver.

The fact that Clemence and Graves carried a great many formal letters of greeting from various organizations and individuals (including Premier John Oliver) is proof of their carefully laid plan to conduct such an historic flight, and confirms the fact that they had been boldly upstaged by Captain Eckley the previous day. There was also a letter from Victoria Postmaster H.F. Bishop to Vancouver Postmaster Macpherson, and this is claimed to be "the first time that any aerial mail has passed between postmasters in Canada" (*Times* 1919.05.14: 7).

There was another significant "first" that occurred shortly after takeoff from Willows:

> After attaining a fair altitude Lieut. Graves let out the wireless aerial and at once called up Sprott-Shaw wireless school. The first message received was as follows: "Aeroplane 'Pathfinder' outward bound for Vancouver 12.25 p.m." This was the first wireless message ever sent from an aeroplane west of Toronto. An instrument of one-quarter kilowatt was installed on the Pathfinder yesterday by the Aerial League and was tested out in a flight late last evening.
>
> —*Times* 1919.05.14: 1

**SUNDAY, MAY 18, 1919**

Lieut. Robert Leslie Rideout and Lieut. W. Harry Brown fly from Willows Aerodrome to Seattle, with an unplanned emergency stop en route at Whidbey Island.

> The Pathfinder, piloted by Lieut. R. Rideout and with Lieut. H. Brown, M.C., as passenger, will set out at 10:30 this morning for Seattle, and will make the trip to Ballard Park as near as the crow flies, as the weather permits. The plane will carry an invitation to Mayor Hansen, of Seattle, from Mayor Porter to attend the 24th of May celebration here. After staying long enough to take on gas and oil, the Pathfinder will start for home.
>
> —*Colonist* 1919.05.18: 7

B.C. Archives D-04019

Robert L. Rideout (left) delivers a letter from Victoria Mayor Robert J. Porter to Seattle Acting Mayor W.D. Lane. W. Harry Brown (right) accompanied Rideout on the pioneering round-trip flight.

Louis Grant and Harry McDiarmid had travelled by ferry to Seattle to prepare the way for their fellow aviators. Events did not turn out exactly as predicted. Adverse winds forced Rideout and Brown to land near Coupeville, on Whidbey Island, but after refuelling they resumed their flight. Escorted by Bill Boeing's own pilot, Eddie Hubbard, they were guided to Jefferson Park Municipal Golf Course in Seattle, where a crowd of about two thousand awaited their arrival. On the golf course, they presented the letter of invitation to Seattle's Acting Mayor Lane (see photograph above), and then had no choice but to stay overnight before undertaking the return flight to Victoria on Monday, May 19.

The aircraft was the original *Pathfinder*, in May 1919 the Victoria Aerial League's only Curtiss JN-4.

For his careful research on this dangerous and historic mission we are indebted to B.C. author and aero-philatelist James A. Brown, in *Hubbard: the Forgotten Boeing Aviator* (1996: 49–53). Jim Brown points out that this flight is the original reason why, for over eighty years, there has always been such a large representation from Washington State in Victoria's May 24th Parade.

We shall see more of Harry Brown in Chapter 6. Like Brown, Grant, and other Victoria pilots—Alex Holden, Harold Bromley, Arthur Raybone, Norman Goddard—Bobbie Rideout sought better opportunties in the United States. For many years he made his living in Hollywood, working as a stunt-man and occasional tough guy for Columbia Pictures (*Vancouver Sun* 1969:03.05: 35).

After he had retired to Vancouver, Rideout presented the B.C. Provincial Archives with documentation of this flight. He died in Vancouver on May 18, 1972, age 74.

Gordon A. Cameron Collection

Probably the first-ever aerial photograph of downtown Victoria, this shot was taken on June 11, 1919, through the rigging of *Pathfinder II*. We are looking west over Victoria Harbour toward the Sooke Hills, from a point above the intersection of Fort and Yates. At the head of View Street we see Girls' and Boys' Central Schools, with the old Fire Hall No. 1 at the corner of Camosun. A camera bracket partly obscures the left foreground. The picture was very likely taken by Gordon Cameron, flying as observer with pilot James Gray.

Because there has been some historical confusion and mistakes in identifying photographs, let us briefly review the chronology of Aerial League activities in Victoria. Those who wish to explore the facts more closely should consult *The Canadian Civil Aircraft Register* (*CCAR 1920–1928*).

Between May 1919 and March 1922, there were probably only two wheel-equipped aircraft based in Victoria, both Curtiss JN-4 Canucks bought in the first instance by the Aerial League of Canada, Victoria Branch, and operated out of Willows Aerodrome. These were given the names *Pathfinder* and *Pathfinder II.* Both were purchased from Ericson Aircraft Ltd., who were the authorized dealers for the surplus military machines.

*Pathfinder* was shipped to Victoria on Monday, April 28, 1919, and assembled in time for its first trial flight on Friday evening, May 2. It was the star attraction of a one-airplane aviation meet staged at Willows on Wednesday, May 7, when numerous photographs were taken. This was the only aircraft flown by Victoria pilots during the next month, which saw a keen rivalry with their Vancouver counterparts and an historic round-trip flight between Victoria and Seattle (May 18–19, 1919). *Pathfinder* was meant to be Victoria's entry in a major air show scheduled for Willows on Saturday, June 11, 1919, but was seriously damaged in Vancouver on the very morning of that event (Corley-Smith 1989: 126). The Victoria flyers borrowed another Jenny in Vancouver, flew it with good results in that day's air show, and then purchased it from Ericson, naming it *Pathfinder II.*

Three aircraft took part in that Victoria air show at Willows on June 11: the future *Pathfinder II* (bearing a large identification number 2 on its fuselage, and displaying an original air-force registration of C-1293 on its rudder); and two Vancouver aircraft flown by Ernest Hoy and George Dixon, respectively numbered 3 and 4 (C-227). *Pathfinder* (C-242) was *hors de combat* in Vancouver. It was repaired to fly again for the Victoria Branch, but the conspicuous events of that summer, including an historic air-mail flight in August between Victoria and Nanaimo, featured *Pathfinder II.*

Victoria teenager Gwendolyn Richards, who named *Pathfinder*, is about to fly at Willows on June 11 with visiting Vancouver Aerial League pilot George Dixon.

In the next chapter, we shall learn that *Pathfinder II* was sold in May 1920 to former Victoria Aerial League pilot Harry Brown. Under the newly adopted national registration system, it was marked as G-CABU, equipped by Yarrows with twin pontoons, and flown in a brief but impressive commercial venture that ended in Brown's heroic survival after the plane was lost at sea, in September 1920. That lineage is very well documented (*CCAR 1920–1928*). Less certain, but highly probable, is that the original *Pathfinder* enjoyed a future life in a different commercial venture as G-CACO, destroyed during a forced landing in suburban Oak Bay on March 19, 1922.

Three Curtiss Canucks took part in the Willows Air Show on June 11, 1919. Two were Vancouver Aerial League aircraft flown by Ernest Hoy and George Dixon; the third (No. 2, in the foreground) would become *Pathfinder II.*

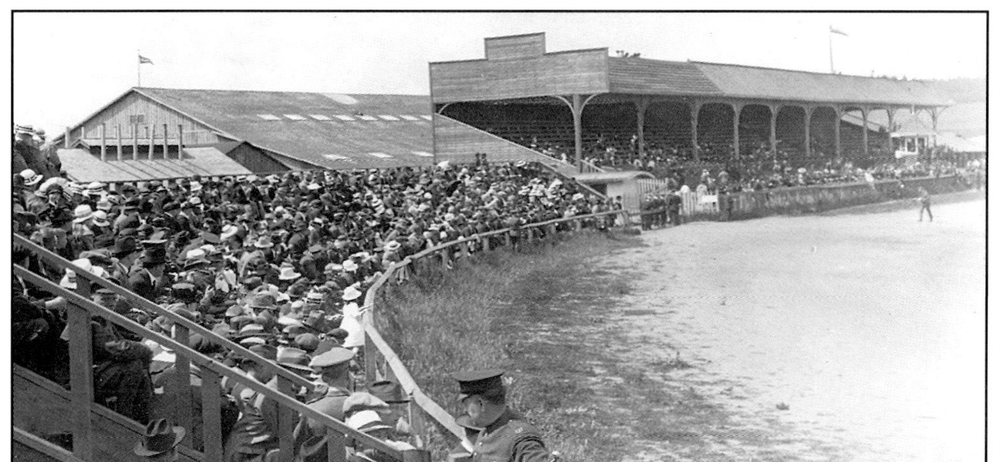

B.C. Archives A-06799

This excellent view of the race track and grandstand at Willows Fairgrounds can be dated to June 11, 1919—the Aerial League Meet advertised at right. There were at least 2,000 in the stands, with many freeloaders parked in cars outside.

*Colonist* advertisement, June 10, 1911, p. 12

For the May 24th holiday, Vancouver had scheduled its first major aviation meet at Minoru Park in Richmond, postponed until May 31. Its most spectacular feature was the innovation of wing-walking and similar acrobatic stunts, performed by a local Vancouver daredevil named J. Harry Fieldhouse. Now it was Victoria's turn, on Wednesday, June 11. Four planes were to arrive at Willows in formation; but their number was reduced to three when *Pathfinder* was damaged in a last-minute accident before leaving Vancouver; the Victoria team was generously lent one of the Vancouver aircraft. Harry Fieldhouse came as a passenger with George Dixon, and performed impressively, to loud acclaim. The other Vancouver pilot was Ernest Hoy, now known for his solo flight to Calgary on August 7, 1919.

At the earlier Vancouver meet, Jack Clemence had ably upheld Victoria's honour, and he gave another brilliant show of stunt flying on June 11. On this occasion, however, the most exciting event was probably a three-way match race to Sidney and back, featuring pilots Hoy and Dixon in Jennys 3 and 4, plus Victoria's Bobbie Rideout in the borrowed Number 2 aircraft. In a thrilling finish, Rideout edged out Hoy and Dixon, who arrived twenty seconds later in a dead heat. This success may have inspired the Victoria Branch to buy that winning aircraft: *Pathfinder II.*

B.C. Archives I-51757

Both photographs above have been misinterpreted. At left, in Curtiss Jenny No, 3 on June 11, 1919, is Vancouver pilot Captain Ernest C. Hoy, D.F.C., about to give a ride to Miss Marie Sweeney of Victoria, who won the prize at auction. Though newspaper reports leave no room here for doubt, the woman has been incorrectly identified as Emily Carr. The other photograph was almost certainly taken at Willows the same day, because the man on the right is again Ernest Hoy. The usually impeccable Frank Ellis identified him many years ago as Alfred Eckley—pilot on the pioneer flight of May 12—and that mistake has persisted (Ellis 1961: 154). With Hoy is presumably his Vancouver passenger on June 11, according to the Victoria papers a well-known former local man, Captain Norman Hall.

# VICTORIA-TO-NANAIMO AIRMAIL

A place in the record book was won by the Victoria duo of Captain Jimmy Gray and Lieutenant Gordon Cameron on Saturday, August 16, 1919, when they flew the first air-mail delivery from Victoria to Nanaimo. This was an experimental, one-time venture, with post-office approval. Leaving Willows at 1:20 p.m. (photo below), *Pathfinder II* reached Nanaimo (Collieries Farm) at 2:15 with one bag of mail containing 96 covers, having flown the 74 miles in 55 minutes. The return flight to Victoria was completed in 50 minutes (7:55 to 8:45 p.m.), but no mail was carried, because of a misunderstanding. (*Colonist* 1919.08.17: 4.)

The star attraction in a sports carnival that evening at Nanaimo Cricket Grounds, Jimmy Gray gave a superb exhibition of trick flying: nose dive, spiral glide, looping the loop, and volplaning. He swooped

> "several times close enough to shake hands with those in the grandstand, but his velocity being about 80 miles an hour did not permit of this friendly greeting. One man in the act of lighting a cigarette had the lighted match blown out by the wind. It was close enough to make all the spectators duck their heads and smell the lubricating oil."
>
> —*Nanaimo Free Press* 1919.08.18: 1

Courtesy of Donald G. Cameron

Born in Calgary in 1896, Gordon Cameron moved as a boy to Victoria, where he enlisted on April 11, 1916. While serving with the 62nd Battery, 25th Artillery Brigade, C.E.F., he was commissioned to the Royal Flying Corps as an Observer in 1917. On June 15, 1918 he was severely wounded over Ypres and invalided back to Canada. In the Aerial League of Canada, Lieutenant Cameron flew often with Captain James Gray.

After being called to the bar, he maintained a lifelong interest in aviation. He was a key figure in B.C. Airways during the late 1920s, and in the revival of the Victoria Flying Club after World War II.

Courtesy of John A. Gray

James Gray was an accomplished pilot who attributed his survival on the Western Front to his experience as a crafty duck-hunter in his native Victoria. His younger brother Robin was not so lucky (see page 20).

Born in 1891, Jimmy Gray earned a B.Sc. degree at Toronto after high school graduation in 1909. Like Gordon Cameron, he transferred from the Canadian Infantry to the Royal Flying Corps in 1917, and flew on operations with the No. 27 Squadron in France until July 1918, earning promotion to Captain.

After flirting with aviation in 1919, he chose a career in marine engineering. After their father's death in 1923, James and his older brother Jack ran Marine Iron Works in Victoria for many years.

During World War II James Gray served with the RCAF as a Link Trainer instructor.

Courtesy of John A. Gray

In both Victoria and Vancouver, the Aerial League seems to have run out of energy after the summer of 1919. Victorians did watch another air show at Willows on Labour Day (September 1, 1919). A local boy named Roland G. Christy gave a demonstration of daring stunts, climbing on the wing of *Pathfinder II* and dangling from the undercarriage. Later in the day, Harry Brown won a match race to Mary Tod Island and back, defeating Captain MacDonald of Vancouver (*Times* 1919.09.02: 9). This was the last hurrah of the Victoria Branch.

One problem was a chronic lack of funds, which could not be significantly remedied by cabaret dances at the Empress Ballroom. More fundamental was the fact that the League had been unable to persuade ordinary citizens of Vancouver and Victoria that there were real commercial opportunities in aviation. There had been occasional moments of success, such as the delivery by air of a 60-pound ($1,000) order for clothier J.N. Harvey on June 27, 1919, when Lieut. Trim flew a Vancouver Curtiss Canuck to Willows (see photo below). "Keeping faith by aeroplane" was how Mr. Harvey described this shipment, said to be the first merchandise ever carried by air in British Columbia (*Colonist* 1919.06.28: 7). The film delivery on June 2 (sidebar left) had dramatized another potential advantage of commercial air service. Perhaps the major problem was the absence of a resourceful entrepreneur.

Whatever the causes, the Aerial League of Canada had become dormant in B.C. by the spring of 1920, and for practical purposes was virtually defunct.

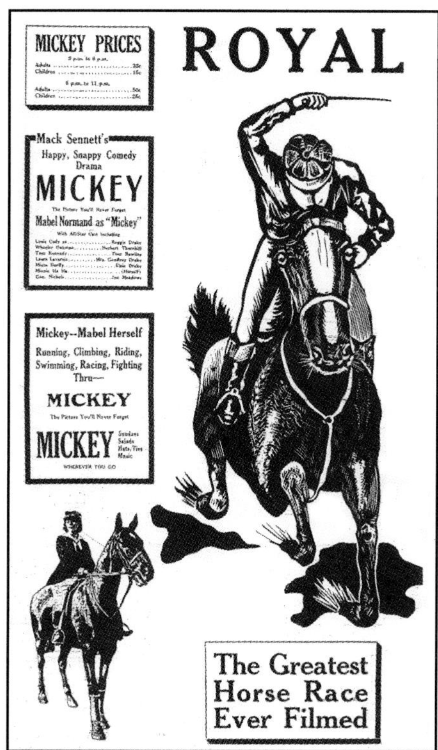


A blockbuster motion picture of 1919 was Mack Sennett's *Mickey*, a horse-racing epic starring the hugely popular Mabel Normand. On Sunday, June 1, the Royal Theatre ran this full-page ad in the *Colonist* to plug its Monday opening, but (disaster!) the film missed the midnight boat from Vancouver. The Aerial League to the rescue! After high-level negotiations, a Vancouver JN-4 flew the film across the Strait on Monday morning, and the 2 p.m. showing went ahead as planned.

Archie Wills Collection

Peter Corley-Smith Collection

The startling picture at right has been identified with the May 1919 visit to Seattle by Rideout and Brown in *Pathfinder*. Finding the scene unnatural, Peter Corley-Smith (1989: 118) wondered if the photograph might have been doctored. Apparently not: the *Victoria Times* ran the same picture on September 28, 1920 (p. 8) as a year-old file photo of *Pathfinder II* "rising from the streets of Port Angeles while piloted by Lt. Jack Clemence." On July 11, 1919, Clemence had been the first pilot to land a plane in Port Angeles, having flown across the Strait of Juan de Fuca with Lt. Louis Grant. Landing in the baseball park, he then gave several courtesy flights and offered a typically flamboyant "stunting exhibition," an event that "caused quite a little excitement." No wonder! (*Times* 1919.07.11: 4)

# 6. Pioneer Local Ventures in Commercial Aviation ▪ 1920–1926

William Henry Brown (always known as Harry) was born in Victoria on March 12, 1894. His father Benjamin William Brown was a fish dealer (1901 census). For many years the family resided at 941 Green Street, near Athletic Park, where Harry had only a short walk to attend Victoria High School. He was a twenty-year-old bank clerk when war was declared in 1914. Already a member of the Canadian Signal Corps militia, he signed up for overseas service on September 23, 1914.

As an infantry soldier, Harry had actually survived the Battle of Vimy Ridge before he transferred to the Royal Flying Corps, where he earned official status as an ace. He was credited with shooting down nine enemy aircraft (Bashow 2000: 185), and was awarded the Military Cross. Returning home as a decorated war hero, Harry became a local celebrity through his exploits as a record-setting pilot with the Victoria Branch of the Aerial League of Canada. But that was merely one of many phases in his eventful life. This young man was a pioneer in British Columbia commercial aviation, flew also for a short time in government service, and enjoyed a highly successful career after moving to California.

### VANCOUVER ISLAND AERIAL SERVICE

In 1920, Harry Brown entered partnership with Norman A. Goddard of Victoria and Courtenay (a strong technician) to form the Vancouver Island Aerial Service. In early May they bought and registered Curtiss JN-4 (G-CABU), the former *Pathfinder II* that had then been owned for eleven months by the Victoria Branch of the Aerial League of Canada (*Times* 1920.05.11: 4).

> The Willows aerodrome will be used as a base for the next month, after which time the machine will be fitted with floats and taken North as a seaplane. The base up the island is to be at Comox, where a site has already been secured, suitable to either type of landing required. . . .
>
> Although open for all varieties of flying enterprise, the new Aerial Service Company, state the promoters, will endeavor to work up a passenger traffic running to the more unaccessible places on the coast to the North. Mr. Brown will pilot the Curtiss . . . .
>
> —*Times* 1920.05.13: 5

Before putting into operation this very adventurous program, Brown and Goddard had flown into several less remote regions on Vancouver Island, including Duncan and the Alberni Valley, which their aircraft was the first to visit in June 1920 (see photo page 35).

(Above) Vancouver Island Aerial Service Jenny G-CABU is seen on the ways of the Royal Victoria Yacht Club, where the first installation of pontoons is being removed, in preparation for a new and improved design.

(Left) Yarrows Shipyard, August 20, 1920. New brass-sheathed pontoons were fitted and tested on G-CABU. "E.W. Izard, the popular superintendent of Yarrows' plant, made the return journey to the Willows beach with Pilot Harry Brown, and was given an opportunity of being one of the first to ride in a converted Curtiss aeroplane-seaplane" (*Times* 1920.08.21: 8).

In the photographs to the left, we see Curtiss Jenny G-CABU (above) at Willows Beach, where an adjustment is being made to the new pontoons; and (below) on the beach at Cadboro Bay, where the seaplane appears to be waiting for a change in the tide.

The new Air Board registration system came into effect in early 1920. Every Canadian civilian aircraft was now registered with an initial "G" (for British) followed, after a hyphen, by "CA" (for Canada) and two more letters allocated in alphabetical sequence. This registration system would prevail until 1929.

Exactly a month after the pontoon conversion, on September 20, 1920, the former *Pathfinder II,* now G-CABU, failed Harry Brown and almost cost him his life. It is an amazing story, recounted in the pages of the *Nanaimo Free Press* on September 25, 1920 and the *Victoria Times* on September 27. By an eerie coincidence, the *Times* ran a photograph of the Yarrows conversion on September 22, when Brown was fighting for survival, his predicament completely unknown to the world.

On the afternoon of Monday, September 20, Brown left Alert Bay at 3:15 on a flight path from Alert Bay to Prince Rupert, planning to stop at Bella Bella for refuelling. Suddenly, near Nalau Island, his engine quit cold. Making a successful forced landing in the fast-approaching darkness, he ran into even more trouble when heavy seas carried his machine onto a reef, from which he managed to clamber ashore with great difficulty. By morning the Jenny was gone, never to be seen again.

Twenty-four hours on an uninhabited island convinced Brown that he would either have to do something decisive or starve to death. Selecting a suitable log, he set out with the hope of paddling to the mainland, or perhaps of being sighted by a passing ship. On Wednesday, after a day and a night astride the log, now well out into Queen Charlotte Sound, he caught the attention of the cannery tender *Hidden Inlet,* bound from Alaska to Seattle. His legs numb and swollen, poor Harry was shivering and exhausted, but still paddling determinedly. On Friday evening, September 24, he was put ashore at Nanaimo.

For those who have any acquaintance with hypothermia in B.C. coastal waters, this must seem a totally miraculous escape from death. The story speaks volumes about a man who had survived Vimy Ridge and the aerial dogfights of World War I. Harry Brown quite rightly received a hero's welcome on his arrival in Victoria.

This was the end, not only of G-CABU, but also of Vancouver Island Aerial Service. Brown now found employment as pilot-navigator on forestry and fisheries patrols from Vancouver's Jericho Beach.

RECOMMENDED READING:
Chris Weicht, "Vancouver Island Aerial Service: First of the Many," *B.C. Aviator* 2.2. (Sept/Oct 1992): 17–19.

## VANCOUVER ISLAND AERIAL TRANSPORT COMPANY

In June 1921 Norman Goddard formed Vancouver Island Aerial Transport Company, using another Jenny that was likely the original Aerial League *Pathfinder*, registered now as G-CACO. This aircraft came to grief on March 19, 1922, when Major Kenneth F. Saunders took off from the Willows Oval with his wife as passenger. As they were climbing for altitude, the engine quit suddenly. While making a very tight forced landing in a field near Dalhousie Street and Cadboro Bay Road, Major Saunders struck a fence and the plane was demolished. Mrs. Saunders, unfortunately, was quite severely injured.

A Victoria native, Ken Saunders would later be one of Canada's more notable airmen, becoming chief pilot for the Fairchild Aerial Survey of Canada. When the Fairchild company was taken over by the new Canadian Airways in 1929, Ken became involved in their bush operations over most of the great Canadian bush country, from Labrador to the Arctic and west to Alberta.

Canadian Airways Bulletin 5.6 (Dec. 1933)

Kenneth F. Saunders

He capped his career with 22 years in the Department of Transport, serving as Superintendent of Air Regulations for the Edmonton district.

## Chinese Commercial Aviation School ▪ May 1922–February 1923

The first commercial flying school in Victoria was formed in May 1922 by a group of wealthy Chinese led by Chan Dun and Lee Quong Yee, with support from the New Republic Printing Company of the city. The school had earlier been established in Saskatoon. Its aim was to train pilots for commercial aviation in China, where the need and opportunties were abundant. There was a strong and unconcealed military motivation, as war clouds were already forming over Asia.

Student pilots of the Chinese Flying School pose in front of G-CACJ, with instructor Harry Brown in the cockpit.

Courtesy of Harry Brown

By now a prominent and highly experienced aviator, W. Harry Brown was hired as chief instructor. Norman A. Goddard rejoined his partner from Vancouver Island Aerial Service as the new School's air engineer.

An 80 h.p. Curtiss JN-4 Canuck (registered as G-CACJ) was purchased from Ericson Aircraft Company in Vancouver, and equipped by Vancouver's Hoffar Shipyards with a Brown-designed single pontoon. By May 1922, a floating hangar had been constructed at Inskip Island at the head of Esquimalt Harbour. With eight students on hand, Harry began daily flights over the harbour area. When the weather conditions were not too kind to the frail Jenny, it was converted to wheels and flown from a large field adjacent to Interurban Road, known as Panama Flats. It was there that a minor accident occurred. While one of the students was pulling the propellor through, the motor sprang to life; failing to get clear in time, he received a nasty thump on the head.

Early the next year, the School was back in Esquimalt Harbour—but not, alas, for very long. On February 24, 1923, on a solo flight, student Hip Quong stalled the machine at a height of 100 feet and caused it to dive into the water. Although the Jenny was completely wrecked, the unfortunate young man was only bruised and cut; but the accident spelt the end of this venture in commercial aviation.

When the School closed, Harry Brown moved to the United States, where he became a prominent figure in aviation on the American west coast. His B.C. partner Norman Goddard gained some notoriety in the California-to-Hawaii air race that was sponsored by the Dole Pineapple Company on August 16, 1927. At the start of this event, flying in *El Encanto* with Lieut. Kenneth Hawkins of San Diego, Goddard crashed on takeoff; both men were lucky to escape with their lives. Harry Brown stayed with the instructional end of aviation, forming first a flying school at Stanford University and then serving for many years with the Curtiss Wright Flying School, as well as becoming involved with a variety of sales agencies and crop-dusting concerns.

After Pearl Harbor and the American entry into World War II, he became a civilian instructor at California Air Academy, where he taught U.S. Air Force cadets. Three of these were Harry's own sons, who later distinguished themselves as flyers in action during the conflict.

Courtesy of Harry Brown

Here the Chinese Aviation School Jenny (G-CACJ) is at Cadboro Bay. The single pontoon is clearly visible; the two small wing floats ensure balance.

These two photographs were taken in May 1926 at different angles from the Point Ellice Bridge, near Rock Bay. They show a visitor, Consolidated Whaling Corporation's "whale spotter" (G-CAFK). A Boeing C-672 (Model 5) seaplane built in 1918 for the U.S. Navy, it was now based at Naden Harbour in the Queen Charlotte Islands. Note the modified rudder.

The top photo provides a fine panorama of Victoria's Upper Harbour, including the almost new Johnson Street Bridge (1924). The CPR steamship at Victoria Machinery Depot is the *Princess Alice*.

This first airplane in the Alberni Valley was the former *Pathfinder II*, now Vancouver Island Aerial Services' G-CABU, owned by W. Harry Brown and Norman Goddard (see p. 32). Some two months after Joseph Clegg took this photograph, the Curtiss JN-4 was fitted with pontoons by Yarrows Shipyard. The aircraft was lost after Harry Brown's forced landing at Nalau Island.

# 7. The Seattle Airmail Connection ▪ 1919–1929

While reporting the Willows aviation meet of June 11, 1919, Victoria newspapers announced that William E. Boeing of Seattle, the famous aircraft manufacturer, would be flying to Victoria by seaplane on Thursday, June 13. His visit was partly to reciprocate the May flight to Seattle by Rideout and Brown, but also to help the Aerial League of Canada promote the cause of aviation in Victoria. On June 13, he landed near the Yacht Club in Cadboro Bay, and then addressed a Rotary Club luncheon meeting at the Empress Hotel.

Mr. Boeing was accompanied by his friend and pilot Eddie Hubbard. Their seaplane was a modified C-700, specially adapted for Mr. Boeing's personal use as Model CL-4S. It was the first aircraft to clear customs at the Port of Victoria. Earlier that spring, on March 3, 1919, the duo of Boeing and Hubbard had made the first international airmail delivery in North America, flying a special shipment of letters from Vancouver to Seattle.

At the Empress, Bill Boeing predicted that Victoria and Seattle would enjoy a special relationship in the years ahead, linked by a common interest in aviation. For over a decade, his words would appear prophetic, as the two cities became far closer than one might think possible in the 21st century. The kinship was perhaps fostered less by aviation than by regular local steamship service and a shared role as terminus for trans-Pacific ocean liners. But it was Eddie Hubbard who pioneered this close laision.

On October 15, 1920, Eddie Hubbard inaugurated the first international airmail service in North America, flying this same seaplane between Victoria and Seattle. Although no longer a regular Boeing employee, he performed this service with Mr. Boeing's collaboration and approval. In November 1920, he began using a second aircraft that had been custom-built for Mr. Boeing, a much larger and sturdier pusher flying boat, Model B-1.

On this page there are four views of the Boeing CL-4S seaplane at Cadboro Bay on June 13, 1919.

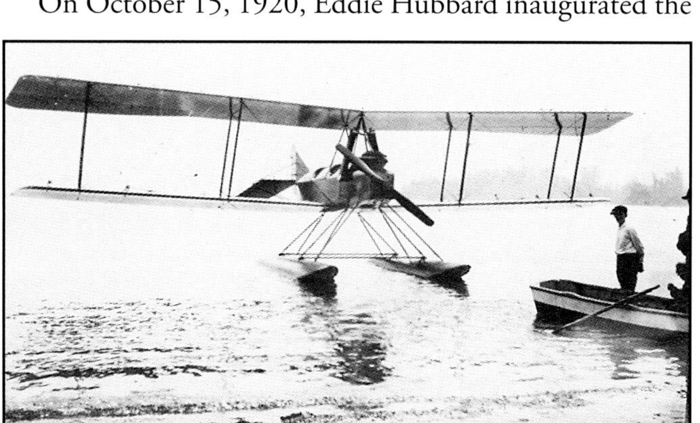

As the American pilot Eddie Hubbard taxied his seaplane up to the JBAA float in Victoria's Inner Harbour, on the afternoon of Friday, October 15, 1920, he earned himself a special niche in aviation history. His round trip that day between Seattle and Victoria would become celebrated as the inaugural flight on the first international airmail service in North America.

More than a year before, on March 3, 1919, Eddie had scored another first, when he and William E. Boeing, his employer and close friend, made the first international airmail delivery in North America by flying a bag of 60 letters from Vancouver to Seattle. But that was a quickly arranged one-shot effort, conceived mainly as a publicity stunt.

In contrast, this 1920 flight brought to reality a carefully planned and highly important service, for which, with Boeing support, Hubbard had won the contract in open competition. The *Victoria Daily Times* editorial to the right (October 4, 1920) explains the rationale of the endeavour. Except for one major interruption in 1923, Eddie flew this mail route between Victoria and Seattle as pilot and entrepreneur from October 1920 to June 1927. He earned an enviable reputation for reliability, becoming a great hero to Victoria children and to aviation enthusiasts on both sides of the border. The Seattle–Victoria service that he pioneered continued until June 30, 1937.

**Victoria Daily Times**

MONDAY, OCTOBER 4, 1920.

Published every afternoon (except Sunday) by

THE TIMES PRINTING & PUBLISHING COMPANY, LIMITED.

Offices: Corner Broad and Fort Streets.
Business Office (Advertising). Phone 1090
Circulation .................... Phone 3345
Editorial Office ............... Phone 45

SUBSCRIPTION RATES:

City Delivery ...........$1.00 per month
By mail (exclusive of city)
Canada and Great Britain ...............$6.00 per annum
To U. S. A. ............$7.00 per annum
To France, Belgium, Greece, etc. ..........$1.00 per month

**THE AIR MAIL SERVICE.**

The object of the service for which Edward Hubbard, the Seattle air pilot, has received a contract from the United States Post Office Department, is to cut down the delay involved in the delivery of American-Oriental mail carried on the ocean liners of the Canadian Pacific Company. At present this mail, if inbound, is taken to Vancouver and then conveyed to Seattle, while outgoing American mail intended for the C. P. liners has to be sent from Seattle to Vancouver. Hubbard's business will be to establish direct connection between Seattle and the Canadian ships at Victoria. He will meet incoming Empresses here, get as much American mail as he can, up to 600 pounds, and take it by plane directly to Seattle, thus saving at least a day. He will bring American mail for the Orient from Seattle to outgoing Canadian steamers at this port, again cutting out the Vancouver relay and saving a good many hours. It is estimated that the service will give Seattle an advantage of twenty-four hours over San Francisco in the delivery of Oriental mail, and in view of the keenness of the competition between the two ports for trade with the Orient this should be an important consideration. Hubbard will receive $200 for each round trip and his contract will extend until June, 1921. The new service will commence on October 15, which also will mark the inauguration of an air service between Key West, Florida, and Havana, Cuba, the two being the first international services in operation out of the United States. It is rather surprising that the Canadian and American Governments have not availed themselves of the opportunity presented by aerial navigation for the delivery of important mail long before this. Is the phenomenon due to the opposition of the agencies which now are in receipt of mail subsidies? In any case it is plain enough that Hubbard's contract is merely the forerunner of what will become a very general business all over the world before long, despite all pressure or opposition exercised at official quarters.

RECOMMENDED READING:
Jim [James A.] Brown. *Hubbard: The Forgotten Boeing Aviator.* Seattle, Portland, Denver, and Vancouver: Peanut Butter Publishing, 1996. The full story of this airmail service is presented in great detail.

(Right) Pilot Eddie Hubbard has climbed onto the wing, while Victoria Postmaster Harry Bishop and Assistant Postmaster George Gardiner stand on the JBAA float.

City of Victoria Archives 98104-23-6848

Victoria Harbour: Friday, October 15, 1920. L–R: Pilot Eddie Hubbard, Assistant Postmaster George Gardiner, and Victoria Postmaster Harry Bishop. We are facing east; the Empress Hotel and Causeway are in the background. On this inaugural occasion, Hubbard brought his 600 pounds of outgoing U.S. mail to this James Bay wharf, where it was taken by truck to the *Africa Maru* at Rithet's Outer Wharf (*Times* 1920.10.15: 12). He then picked up local Victoria mail for delivery to Seattle.

On later flights to Victoria, Hubbard's usual points of transfer were either Rithet's Outer Wharf or the Brackman-Ker Wharf at Shoal Point.

VICTORIA DAILY TIMES, WEDNESDAY, DECEMBER 8, 1920

UNITED STATES AERIAL MAIL SEAPLANE



Already well-known and well liked in Victoria, Eddie Hubbard became the darling of the local press. Editorial writers praised his admirable skill and total reliability. On Saturday, September 3, 1921, the *Victoria Daily Times* devoted a full page to a glowing report on his first year's activity. The headline read "AERIAL POST IS COMMERCIAL SUCCESS: Over Million Letters Speed Through Air in Seattle–Victoria Flights." Some wondered why the Canadian Government could not introduce a similar airmail service between Victoria and Vancouver.

Elwood White took this picture of Hubbard's Boeing B-1 Flying Boat more than a decade after it had been withdrawn from service, then semi-derelict and exposed to the elements at Boeing Field. Soon afterwards, it was completely restored for display in Seattle's Museum of Science and Industry. It weighed about a ton and a half when loaded, flew at 95 mph, and landed at 50 mph. It was initially powered with a 200 hp Hall-Scott engine, later replaced with a 400 hp Liberty. Since the plane was a three-seater, Eddie would often carry a passenger in a hurry to get to Seattle. If the mail shipment was heavy, the passenger and Eddie would be amply surrounded by mailbags, often requiring them to sit high up and exposed to the elements.

# SEATTLE FLYING BOAT CRASHES IN THE STRAIT

March 29, 1923: Its hull grotesquely twisted, Hubbard's Boeing B-1 is stored temporarily at VMD, awaiting shipment by scow to Seattle for repairs.

PILOT OF SEATTLE PLANE INJURED; EDDIE HUBBARD

AERIAL POST PLANE TAKES DIVE INTO WATERS OF STRAIT

Eddie Hubbard Is Rescued Off Brotchie Ledge

Had Just Left With Oriental Mail From President Grant

Eddie Hubbard's only major accident in his seven years on the Victoria–Seattle route occurred on March 29, 1923, just moments after he took off for Seattle from Rithet's Outer Wharf.

A broken rudder control caused Hubbard to plunge into the sea near Brotchie Ledge, within full sight of passengers aboard the *President Grant*. Eddie customarily flew very close to the water—sometimes even taxiing on the surface—thereby conserving fuel and avoiding fog. For this reason, he was not seriously injured in the crash, but only a quick rescue operation saved him from drowning, and his flying boat was heavily damaged (upper photo left).

A less dramatic mishap occurred in October 1926, when fog forced substitute pilot Percy Barnes to land at Cadboro Bay. While proceeding to shore, he struck a deadhead, putting a hole in the hull (Brown 1996: 115). This damage could be repaired in Victoria; in the photo at left, we see N-CADS propped up at the foot of Cormorant Street. Two days later, Barnes was on his way to Seattle.

**MAIL PLANE CRASHES IN CITY**

Huge headline in *Victoria Times* (1926.10.26:1)

On October 26, 1926, another substitute pilot, Gerald Smith, demolished his own Aeromarine flying boat while on a Seattle airmail run. In taking off from Victoria's Inner Harbour at 9:04 a.m., with one sack of mail, Smith lost control of his plane and crashed into the home of Senator Robert F. Green, 502 Rupert Street (right). Senator Green and his wife were in the house at the time, but were not hurt; he was among those who pulled the injured pilot from the wreckage. Miraculously, Smith suffered only cuts and a broken ankle; he soon returned to Seattle on a stretcher. It was found that a light mooring line had wound itself around the propeller hub, causing loss of full power at takeoff.

L–R: mechanic (probably Jack Halloran), pilot Anscel Eckmann, Sgt. R. Owens (B.C. Provincial Police)
Photograph by Associated Screen News Limited, Empress Hotel, Victoria, B.C.

In a period of just over two years after winning the Seattle airmail contract, Eddie Hubbard performed magnificently, flying 35,200 miles in 220 round trips (Brown 1996: 113). Therefore it was an unexpected shock when he lost the contract for 1923–24 to Alaska Airways, which began using a Curtiss HS-2L flying boat. Their pilot on the Victoria run was a young University of Washington law graduate named Anscel Eckmann, a Seattle native who would later have other associations with aviation in Victoria.

The smiling Eckmann appears (left) in a 1929 Victoria celebrity news photo, taken probably on his return from a historic non-stop flight from Seattle to Juneau. He would enjoy a dazzling flying career on the Pacific coast, retiring unscathed at a relatively youthful age.

In transporting the Victoria–Seattle airmail, however, Eckmann and his Curtiss flying boat were no match for Hubbard and his efficient Boeing B-1. In the middle of its one-year contract for 1923–24, Alaska Airways chose to cut its losses by asking Eddie to resume the run on subcontract. By the time he retired in 1927, Hubbard had worn out six engines and flown some 350,000 miles. By then a wealthy man at age 39, he died suddenly after an operation for a stomach ulcer in December 1928. The news was received with shock and sorrow in Victoria.

In the centre above and wearing the same cap in the photo at right is Herold (correct spelling) Walker, who followed Eddie Hubbard as Victoria–Seattle airmail pilot in the year 1927–28. Flying for Northwest Air Services, Inc., of Renton, Washington, he still used Hubbard's original Boeing B-1 flying boat, which was then in its last year of service. The well-dressed men in hats are Canadian postal officials who are supervising the transfer of a very important Japanese document that is being sent via Seattle to Washington, D.C. This item had been accompanied since leaving Tokyo by a Japanese emissary, who appears in both pictures. The locale is the Brackman-Ker wharf, Shoal Point (p. 43).

Victoria's own Brackman & Ker Company had won first prize for its rolled oats at the Chicago World's Fair of 1893. Here we see that Company's wharf at Shoal Point.

Different Boeing models appear in these pictures. Despite its poor quality, the top photo is noteworthy, as it appears to provide a glimpse of a second flying boat at the far right. The aircraft fully visible carries registration NC-115E, which identifies it as Boeing Model B-1E, one of five such aircraft manufactured in 1928. It was not necessarily a Seattle airmail plane.

The photos above and below depict Boeing flying boats against the background of the Brackman & Ker (B&K) Milling Company wharf at Shoal Point, about which more is said on page 43. Victoria aviation historians soon become familiar with B&K Wheat Flakes and B&K Rolled Oats.

The flying boat for which Percy Barnes became best known on the Victoria–Seattle airmail run in the era of Barnes and Gorst Air Line Inc. (1929–1934) was called *Zephyr*, a name that appears on the hull in front of "U.S. MAIL" in the photograph at left. A pusher with an enclosed cockpit, like Model B-1E, it was a unique Model 204A (1929), with registration 875E. Some sources state that it was often referred to as the "Red Boat" because of the colour that it was painted.

The two photographs above are similar views of Percy Barnes's lakeshore home in Tacoma, Washington. The Boeing *Zephyr* (Model 204A) is the aircraft in the right photograph, and the nearer aircraft in the left. Beside it in the left photograph is a similar but different machine; Percy Barnes may have had more than one "Red Boat."

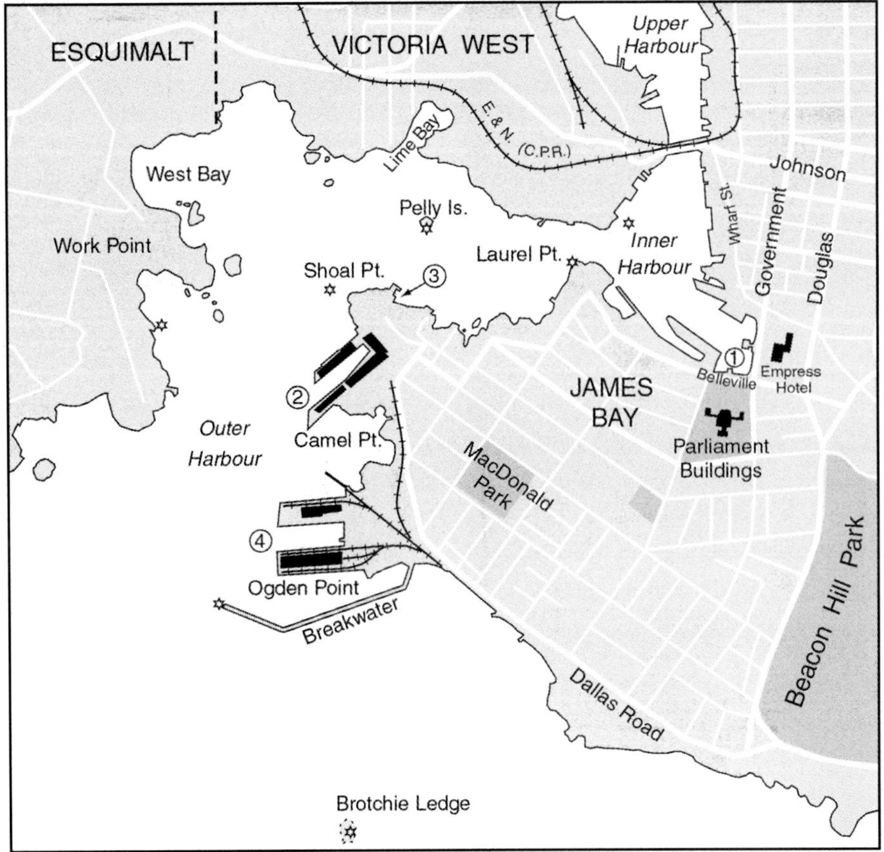


**SEAPLANE DOCKING SITES IN VICTORIA HARBOUR (Map numbers match numbers below.)**

There is a problem of nomenclature. Most Victorians think of the "Inner Harbour" as the docking area near the Parliament Buildings —strictly speaking, "James Bay." Victoria's "Inner Harbour" actually begins at Shoal Point, though modern harbourmasters have introduced the useful new term "Middle Harbour."

① **Victoria Inner Harbour: James Bay**
Here Eddie Hubbard made his historic first airmail pick-up on October 15, 1920. Only for a short time in 1931 was James Bay used for regular passenger service to Vancouver or Seattle.

② **Rithet's Outer Wharf**
In World War II, this area became Victoria Machinery Depot (VMD), and still later, Canada Coast Guard. Here Hubbard often exchanged mail bags directly with trans-Pacific liners. R.P. Rithet's Outer Wharf appears in the photo to the left.

③ **Brackman & Ker (B&K) Wharf**
(See across page). Originally a simple float built by Eddie Hubbard at the northern corner of Shoal Point, this was another favoured location for the exchange of airmail, and later became an air passenger wharf.

④ **Ogden Point Wharf**
Not to be confused with what was then called the "Outer Wharf," this relatively modern pier was rarely used by passenger liners in the 1920s and 1930s, and for that reason seldom visited by the Seattle U.S. Mail pilots.

City of Victoria Archives 98110-08-7541

The photograph above of R.P. Rithet's Outer Wharf was taken about 1920, and therefore is the view that Eddie Hubbard (or a trans-Pacific steamship traveller) would have enjoyed on arrival in Victoria. Brackman & Ker National Mills is the tall narrow structure on Shoal Point, in the distance, left.

The picture to the right shows Hubbard's Boeing B-1 flying boat making a direct exchange with the tender from an ocean liner. The back of the photograph is labelled "Ogden Point," but it seems more likely to have been taken at Rithet's Outer Wharf.

## VICTORIA'S B&K WHARF

In this archival photograph from the early 20th century, we see Brackman & Ker's National Mills from the opposite (Inner Harbour) side of Shoal Point, looking roughly southwest. Today, it would be like viewing HMCS *Malahat* from Fisherman's Wharf. On what appears to be a gentle slope—can that be sand where the barge has been beached?— Eddie Hubbard built a simple wooden floating wharf when he began flying to Victoria in 1920.

That so-called B&K Wharf is partly visible in the two lower photographs, taken from opposite directions in 1928 or 1929. Both show a Loening amphibian Air Yacht (NC-9156) from the fleet of Gorst Air Transport, an enterprising Seattle company operated by Vern C. Gorst, an older man who often acted as his own pilot.

In 1928, Vern Gorst entered partnership with Seattle pilot Percy Barnes; Barnes and Gorst received the Victoria airmail contracts for 1928–29 and 1929–30, trying also other types of flying service, including (briefly) a scheduled passenger run. Their Seattle base was at Lake Union.

In May 1930, Barnes and Gorst Air Line Inc. were awarded an extension of their U.S. airmail contract to June 30,1934. (*Times* 1930.05.06: 15)

B.C. Archives C-08963

Photo by Lloyd Jarman

Vern C. Gorst (C. for Centennial— born in 1876). Gorst Air Transport linked Victoria and Seattle.

In partnership with Vern Gorst through the 1930s, Percy Barnes apparently carried on as Victoria–Seattle airmail pilot until the route came to an end in June 1937. The great Trans-Pacific ocean liners were gradually withdrawn from service, and Victoria lost its strategic importance as a transfer point. Also, both United Airlines and Canadian Airways had introduced a Vancouver–Seattle airmail service. The photographs in this chapter likely extend only to 1929. The three on this page, all from roughly the same period, show Barnes with a Boeing flying boat, probably *Zephyr* (875E), at the flimsy old B&K wooden float. In each case, we are looking north across Victoria Harbour to the Songhees Industrial Reserve.

Lean and lanky, Percy Barnes was always elegantly dressed. Like Eddie Hubbard, he was willing to carry a passenger or two along with the mail.

The background details across the harbour are revealing. In the top photo, the CPR coast steamship is either the *Princess Kathleen* or the original *Marguerite*. Commercial buildings on the Victoria West shoreline then extended from Canadian Western Cooperage near Lime Bay (formerly Sweeney Cooperage, centre of bottom photo) to Sidney Roofing Company, located between Songhees Point and the Johnson Street Bridge.

PART THREE

# A MUNICIPAL AIRPORT FOR VICTORIA?

## 1927–1933

**Vote For the Airport By-law Monday June 29**

The Airport Is a Greater Victoria Project

Saanich and Oak Bay are willing to contribute their share to the cost.

## 8. B.C. Airways and Victoria Aero Club at Lansdowne ▪ 1927–29

Ted Cressy Collection

Lansdowne Field, 1928: B.C. Airways' Alexander Eaglerock (G-CAIS) is about to land on the main runway. The aircraft is approximately above the spot marked "X" on the map—the modern Aldridge Street. In 1928 there was almost no residential development between Lansdowne Road and University School (see map page 65).

The shaded area on this map identifies Lansdowne Field, a 77-acre (31-hectare) property leased by British Columbia Airways of Victoria, opened as an "aerodrome" on January 28, 1928. Comprising all the so-called Saanich Panhandle west of Richmond and south of Lansdowne, it measured on average almost exactly 400m wide and 800m long (roughly a quarter mile by a half mile). The airfield was separated from Shelbourne by a narrow strip of City-owned property, and extended south to Sixth Street, now Newton.

The original runways were little more than well-worn tracks in the grass. However, they are clearly visible in an October 1928 RCAF aerial photograph, as are the two hangars near Lansdowne Road, and also the two oak trees that gave student pilots fits. The north–south runway must have been judged long enough to meet the commercial licensing standard of 1,800 feet. It was proposed to extend this main runway to 2,400 feet, almost the full length of the field. That would have meant enclosing Bowker Creek, which was partially covered by heavy wooden planks during the B.C. Airways era.

White lines locate the roads of the subdivision built in the 1950s; the broken white line marks the fence that now separates it on the north from the spacious field owned by School District 61. The two modern buildings shown in white at the northeast corner are Lansdowne Middle School and the Girls' Alternative Program. Near each of these is a cairn marking an historic moment in Canadian aviation.

Cairn 1 (near the G.A.P. school) honours the achievements of William Wallace Gibson, who flew his Twin-plane here in September 1910. Cairn 2, in front of Lansdowne Middle School, celebrates the visionary endeavours of B.C. Airways, whose ill-fated dream of soaring to national prominence is told in this chapter.

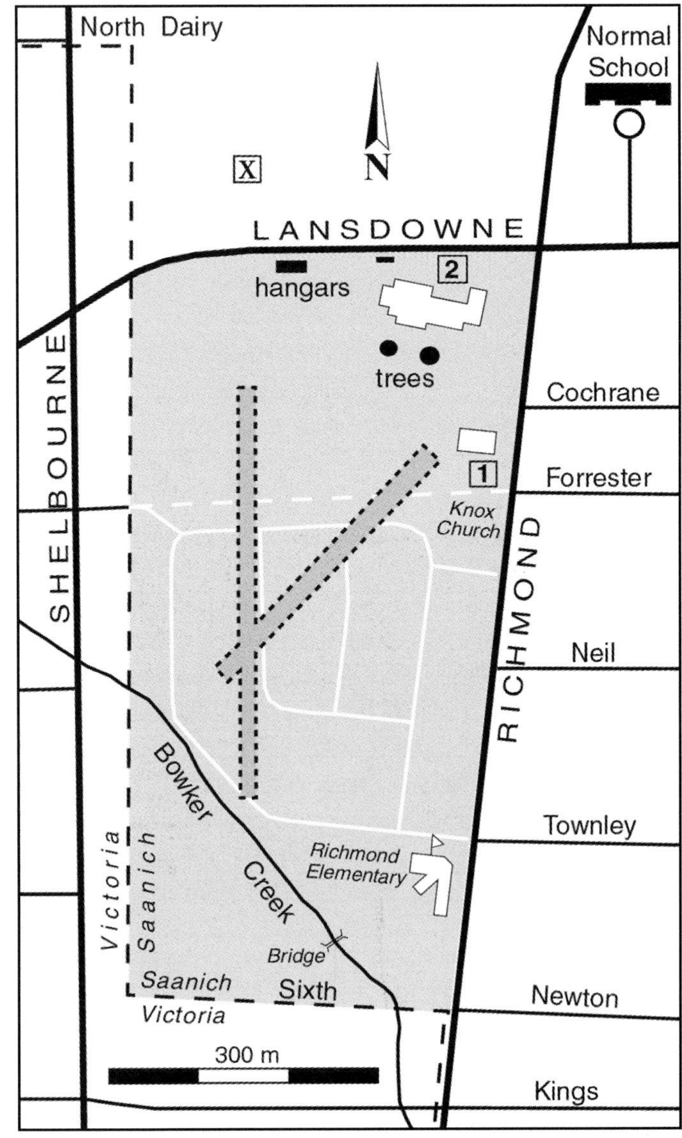


The spacious field was sometimes called Deans Farm, after pioneer settlers George and James Deans, who had both established farms along Richmond Road in 1858.

Ted Cressy Collection

Lansdowne Field, 1928. Here the same aircraft, B.C. Airways' Alexander Eaglerock (G-CAIS) is above the main north–south runway of Lansdowne Field, apparently about to touch down. The three spectators seem unconcerned. As we look northeast towards the Provincial Normal School, we see the still unfinished Victoria Aero Club hangar, built in 1928.

Despite some competition from the likes of Babe Ruth and Lou Gehrig, the pre-eminent superstar of 1927 was the charismatic young American, Charles Lindbergh. From a global perspective, Charles Augustus Lindbergh was a hero in a class by himself—an overnight celebrity such as the world has seldom witnessed before or since. His trans-Atlantic solo flight in *The Spirit of St. Louis* from New York to Paris, on May 20, 1927, had an electrifying effect on the human imagination, an event that bears comparison with the Apollo 11 moon landing of July 20, 1969.

The world would never again be the same for young dreamers like Victoria's 16-year-old Maurice McGregor, whose sights were now set firmly on the stars. Brothers Ernest and Cecil Eve, creative entrepreneurs who owned a local automobile franchise, glimpsed the dawning of a new age in commercial transportation. Thanks to the Eve brothers and their eager supporters, Victoria's Lansdowne Field was developed in 1927 and officially opened in 1928 as the home base of British Columbia Airways, Limited. Over the next four years it would witness many a triumph and a disastrously crushing tragedy. During its period of decline, in the pre-dawn hours of October 22, 1931, it celebrated the arrival of the great man himself—the 29-year-old Charles Lindbergh and his new young wife Anne Morrow, who took off from this Victoria airfield on a flight path to the eastern U.S.A.

The local impact of Lindbergh's solo flight to Paris can be easily gauged by skimming Victoria newspapers during the second half of the 1920s. Prior to May 1927, aviation stories were usually consigned to the inside pages, or given only minor coverage on page one. Granted, there had been a flurry of excitement during the heady but now-distant era of the Aerial League of Canada and the early flights of Eddie Hubbard between Victoria and Seattle; but nothing much had been happening locally since Harry Brown headed south to the U.S.A. After Charles Lindbergh, however, there was a steady succession of screaming banner headlines in both the *Times* and the *Colonist*. Readers were hungry to read about every pioneer flight or competitive air race anywhere in the world. Once B.C. Airways began flying from Lansdowne Field to Vancouver and Seattle, its operations were covered with breathless intensity. When its Ford Trimotor crashed into the sea on August 25, 1928, the whole community grieved a devastating loss.

It was early in the summer of 1927 that Victoria first heard about a new initiative in aviation. Newspapers reported an evening meeting on July 11 in the office of Eve Brothers, Ltd., well-known automobile dealers who held the Graham-Paige franchise. J. Ermest Eve was then 39, and his brother Cecil H. Eve was 36. Neither had any hands-on experience with flying, but they were good businessmen who shared a love of machines and bold ventures.

J.E. (Ernie) Eve
Drawing by Mort Graham, *Colonist Islander* 1955.12.04: 1

Those who met that night discussed three different but closely related strategies. One aim was to secure the Victoria agency for a light aircraft of the D.H. Moth type. A second goal was to buy a larger, modern aircraft suitable for freight and passenger service. A third was to form a local aero club that might attract veteran flyers from the Great War as well as new and younger enthusiasts from the community. Former aviators who attended that meeting included Gordon A. Cameron, an active member of the old Aerial League of Canada who was now becoming a well-established Victoria lawyer.

During the second half of July 1927, Ernie Eve travelled to Cleveland and Detroit to visit his usual automotive contacts but also to explore new possibilities in aviation. While holding the fort in Victoria, Cecil expressed the opinion that he and his brother were looking primarily at flying boats, large and small. "Mr. C.H. Eve declared that his firm was not interested in land planes since landing fields are scarce on the Coast." (*Colonist* 1927.07.16: 5). This position is not surprising, given the fact that the City of Victoria had been actively exploring for six months the possibility of developing an air harbour at Esquimalt or Victoria's Inner Harbour. Meanwhile, however, brother Ernie was undergoing a spiritual conversion, having been privileged in Detroit to speak with Edsel Ford and the mighty Henry Ford himself. He heard first-hand how the Ford Motor Company was in the process of adapting Anthony Fokker's successful design for a three-engine passenger aircraft into a large, all-metal Ford-Stout Trimotor that would clearly be state of the art. Ernie also visited several plants that specialized in small aircraft.

While Ernie Eve was still en route homeward from Michigan, brother Cecil announced on July 29 that two planes of the land type had been ordered, and that the services of an excellent pilot and an experienced mechanic had been obtained. These were A. Haliburton (Hal) Wilson and T.H. (Ted) Cressy, both young English World War I veterans who had recently arrived in Canada. These two men now become figures of central importance in our narrative. Although the Eve brothers were novices in a complex venture, they had the savvy to surround themselves with experienced professionals. They could have found none better than Wilson and Cressy, who both left a major imprint on Canadian aviation.

Canada's Aviation Hall of Fame. Portrait by Irma Coucill

A.H. (Hal) Wilson

On August 5, 1927, another meeting in the Eve Brothers' office led to the formation of the Victoria Aero Club, a short-lived organization replaced by a different type of Aero Club just six months later. For the moment, Ernest Eve was president; Lieut. Hal Wilson, vice-president; J.W. Cox, secretary; and Gordon Cameron, legal advisor (*Times* 1927.08.06: 9). Ernie Eve now began promoting aviation in speeches to groups like the Vancouver Board of Trade, while the Victoria community sensed that grand events were about to unfold.

That same month a possible competitor came into view when Dominion Airways, Vancouver, revealed its intent to serve Victoria with a de Havilland Moth, stationed likely at Esquimalt Harbour. Ted Dobbin was to be pilot in charge of the operation, but this was one of several prematurely announced plans for Vancouver–Victoria service that never came to fruition.

Ted Cressy Collection

T.H. (Ted) Cressy

There were other professionals urging the use of water-based aircraft. In September, Pilot Officer W.D. Van Vliet (Jericho Beach RCAF Station), acting for the Department of National Defence, declared Esquimalt Harbour "100 per cent perfect" for a civic air base. In November, World War I ace Major Donald R. MacLaren, then of Pacific Airways, underscored the same message:

> While Vancouver must provide a more elaborate airport to care for both land and water machines, the problem in Victoria is much more simple. Here only water machines must be provided for. Victoria, being on an island, is essentially a place for water machines, as there is no place for land machines to fly to.
>
> —*Times* 1927.11.03: 1

By the end of October, however, the die was cast: land machines it would be. Ernest Eve and his associates—soon to become B.C. Airways—had made firm plans to lease the 77-acre Lansdowne property from its owner, Mrs. Elizabeth Scott. City of Victoria files reveal that the original lease was for a five-year term, from November 1, 1927 to October 31, 1932, at a rental of $162.50 per quarter, or $650 per annum. After the Ford Trimotor crash and the subsequent problems of B.C. Airways, the company asked the City to take over the lease in October 1929, when the contract still had three more years to run. The original lease had included an option to purchase, which B.C. Airways never exercised.

Ted Cressy Collection

Above, as we look to the southeast, we can see almost the full expanse of Lansdowne Field, although the main runway is just out of sight on the right. On the distant horizon, to the right, are houses on the Victoria side of Sixth Street, now an extension of Newton Street. The houses to the left are in Saanich, on the east side of Richmond Road. The aircraft include the first two B.C. Airways biplanes, Driggs Dart G-CAIR and Alexander Eaglerock G-CAIS; to the left is an American visitor, Waco GXE, N-5242.

That top photo was taken from the roof of the B.C. Airways hangar, as we can see from the middle picture, shot from the opposite direction but still showing the Driggs Dart and the Waco in the same position. This hangar, completed in 1927, was built very close to Lansdowne Road.

The bottom photo of the spacious hangar interior may actually date from 1927; the little Driggs Dart appears in the far corner, painted in a different colour scheme from its usual 1928 livery. The top two photographs can be confidently dated to 1928; all three came originally from the Ted Cressy collection, generously shared with Elwood White many decades ago.

Ted Cressy Collection

Ted Cressy Collection

In late October, the foundation of a newly constituted Aero Club had been laid when the Federal Government announced a program to make available one or more free de Havilland Moths, if certain conditions were met. This program was explained in Victoria by Major-General J.H. MacBrien, an occasional visitor from Ottawa. The new Victoria Aero Club had a close relationship with B.C. Airways; but because it was a distinct, non-commercial operation, we shall treat it separately.

The first newspaper mention of British Columbia Airways Limited appeared in the *Victoria Times* on November 17, 1927, when the incorporation of that company was announced under a front-page banner headline: AERIAL PASSENGER SERVICE FOR VICTORIA. The report was rather garbled. The Ford-Stout Trimotor, here described as "a stout all-metal machine," was wrongly identified twice as a seaplane, but the story did close with a mention of the new airfield, "a big aerodrome on Landsdown [can anyone spell that name?] Road, near the Willows Exhibition grounds." Both local newspapers correctly identified B.C. Airways' powerful new board of directors: in addition to C.H. and J.E. Eve, members were J.W. Spencer and R.H.B. Ker, with Gordon Cameron as solicitor. This did not sound like a casual, fly-by-night operation.

The very next day, on November 18, 1927, the Company's first aircraft arrived in town (unassembled). This was a Driggs Dart two-seater biplane, soon to be registered as G-CAIR. Hal Wilson and Ted Cressy now set to work to complete the new hangar and make the plane airworthy. Before the end of the year, the Lansdowne property had

been licensed as a landing field, and would be licensed as a commercial airport on January 1, 1928. Certificates of airworthiness were issued to the Driggs Dart and its new companion, an Alexander Eaglerock three-seater registered as G-CAIS. All such official acts of approval—in theory a responsibility of the Federal Department of Civil Aviation—were handled by an inspector from the Jericho Beach RCAF Station in Vancouver, usually Flying Officer A.H. Hull.

Despite poor weather conditions, a successful opening ceremony was held on the afternoon of Saturday, January 28. A vivid account of the event appeared in Sunday's *Colonist*, including quoted remarks by Victoria's Acting Mayor William Marchant, Reeve E.C. Hayward of Oak Bay, Reeve Elrick of Esquimalt, and Reeve Crouch of Saanich. Mr. Crouch drew chuckles by expressing his pleasure that the aerodrome had been located in Saanich, "undoubtedly the finest part of Vancouver Island." Yet this symbolic display of the community's fragmented political structure might have caused a quiver of concern to those who were hoping for the coherent and orderly development of aviation in Greater Victoria.

Ted Cressy Collection

Although this interesting picture cannot have been taken on the day of the opening ceremony, it appears to show Miss Nan Eve admiring the Driggs Dart aircraft named after her. The young man on the far left is likely a student pilot, perhaps W.A. Hughes. Mechanic Ted Cressy is in the checked shirt and B.C. Airways president Ernie Eve in the flying suit. The sergeant in uniform is probably on official business, but his identity is unknown.

The little Driggs Dart II (*Miss Nan*) poses proudly for the camera outside the B.C. Airways hangar on Lansdowne Road. Registered as G-CAIR on December 13, 1927, she was powered by a 40 hp Anzani motor. Her life was short: sold in October 1928 to Sprott-Shaw School of Aviation, Vancouver, she was damaged beyond repair on the day following delivery. Note the contrast in her colour scheme with the picture on the previous page.

Over the next four years, local citizens were surely puzzled by the bizarre spectacle of the City trying to operate and even purchase two separate air fields that were both located entirely in Saanich. It still owned a former air field in Oak Bay.

Still, the ceremony was clearly felt to be auspicious. A crowd of three or four hundred huddled around the hangar to watch B.C. Airways' two aircraft get christened by Miss Nan Eve, "blue-eyed daughter of Mr. Ernest Eve, and, at sixteen, soon to become Canada's youngest girl pilot" (apparently a goal she never attained):

> Garbed in riding costume with blue "slicker," helmet and goggles suggesting the needs of flying, the pretty bobbed hair schoolgirl broke a bottle of champagne over the Driggs Dart II, christening it the "Miss Nan." A few moments later she moved to the larger plane, the Alexander "Eagle Rock" and bestowed upon it the name "Silver Eagle," together with another shower of champagne.
>
> —*Colonist* 1928.01.29: 2

With Flying Officer Hull looking on, we hope approvingly, company pilot Hal Wilson took both *Silver Eagle* and *Miss Nan* aloft in turn, but weather conditions did not allow the special courtesy flights that had been planned.

Now that the airfield was in business, B.C. Airways faced a variety of challenges. Wisely postponing the inauguration of commercial transport until the summer, the Directors now announced an urgently needed sale of shares in the company, and began to promote that cause vigorously. Meanwhile, Hal Wilson and Ted Cressy had already launched the instructional program that was an important part of the overall plan. Fourteen members were signed up for the first course offered by B.C. Airways Flying School. Hal Wilson, of course, was the flight instructor, while Ted Cressy as air engineer gave lessons in mechanics and rigging. Many Victoria boys began their flying careers at Lansdowne Field. The famous included Maurice McGregor, Humphrey (Hump) Madden, and Peter Redpath, all future professionals.

## COINCIDENCE OR DESTINY?

On January 30, 1928—one day after the official opening of Lansdowne Field—its value was dramatically proven when the Seattle Boeing flying boat broke its propeller taking off from Victoria's Inner Harbour while carrying a load of mail from the American liner *President Jefferson*. Because a piece of the propeller smashed one of the plane's ailerons, serious repair work was needed. Returning to the jetty, the pilot asked the newly formed B.C. Airways for mechanical assistance, and phoned Seattle to inform his company, Northwest Air Services, of his predicament. Knowing about Lansdowne Field, he suggested that an available land plane be sent over to Victoria, along with U.S. air mail for the *President Taft*, outbound for the Orient that same afternoon.

His call brought company president William Strain from Seattle to Victoria post haste; he landed his three-passenger Travel Air at Lansdowne as the aerodrome's first intercity visitor. Successfully coping with the mail emergency, and given assurances that B.C. Airways mechanics could repair the flying boat, he returned to Seattle the next morning.

There is a fascinating aspect to this incident. That highly experienced pilot of the Seattle flying boat, then employed by Northwest Air Services, was none other than Herold Walker, who would be hired in July as a senior pilot for B.C. Airways. He would be accompanied by his former boss Mr. Strain when he flew the Ford Trimotor west from Michigan to Victoria. And he would be the pilot of the Trimotor when it crashed disastrously on August 25, 1928.

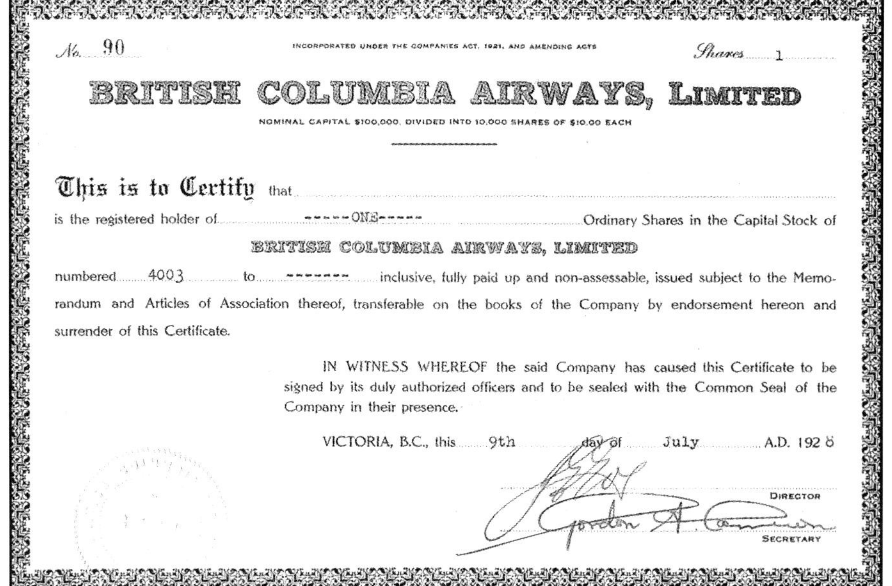
No. 90 INCORPORATED UNDER THE COMPANIES ACT, 1921, AND AMENDING ACTS Shares 1

BRITISH COLUMBIA AIRWAYS, LIMITED

NOMINAL CAPITAL $100,000, DIVIDED INTO 10,000 SHARES OF $10.00 EACH

This is to Certify that

is the registered holder of -----ONE----- Ordinary Shares in the Capital Stock of

BRITISH COLUMBIA AIRWAYS, LIMITED

numbered 4003 to ------- inclusive, fully paid up and non-assessable, issued subject to the Memorandum and Articles of Association thereof, transferable on the books of the Company by endorsement hereon and surrender of this Certificate.

IN WITNESS WHEREOF the said Company has caused this Certificate to be signed by its duly authorized officers and to be sealed with the Common Seal of the Company in their presence.

VICTORIA, B.C., this 9th day of July A.D. 1928

DIRECTOR

SECRETARY

"Mr. E. Eve was with me in the Driggs Dart and, after landing in the field in front of Hatley Park (now Royal Roads), we were successful in selling to the Dunsmuirs $500.00 worth of stock in B.C. Airways Ltd. Due to exceedingly poor judgment I struck a fence in taking off, doing damage to the aircraft amounting to $460.00. After adding to this the cost of transporting the wreck back to Victoria, the net profit of the operation was zero."

—Hal Wilson (March 2, 1928)

Ted Cressy Collection

The original stock certificate is from the collection of Gordon Cameron. In the sidebar above, Hal Wilson describes an almost disastrous fundraising flight in Driggs Dart G-CAIR. This memoir appeared in the *Canadian Aviation Historical Society Journal* 3.3 (Fall 1965): 68.

The little Dart became a stage prop for promoting the cause when it was entered in the Victoria Day Parade on May 24, 1928, urging citizens to buy stock in the new B.C. Airways. We see it here leaving the airfield and headed west along Lansdowne.

Ted Cressy Collection

**A PROFITABLE JULY 1st**
**Hal Wilson Describes**
**Fund-raising in Comox**

"One of the most successful barnstorming days was carried out in a hay field, half-way between what is now the Comox Airport and the Village of Courtenay when, on July 1, 1928, 103 passengers were carried, bringing in a total revenue of $425.00. This, insofar as I am concerned, has never been exceeded. From a flying point of view this consisted of six hours flying—a tiring ordeal when one takes into account the fact that the field measured 1,200 feet from corner to corner.

"During this period two things happened. First, the propeller was broken by a rock. It was repaired by a local carpenter whose wife and family had to be given a free ride as compensation. Second, the tail skid broke. This had to be repaired by the local blacksmith who likewise was given a free hop. In addition to them, the owner of the land and his nine children were all circled around the field. An unpleasant incident occurred on one of these flights when it was discovered, after getting several hundred feet in the air, that one of the passengers was drunk. Only by rocking the aircraft violently could the man be confined to the cockpit by centrifugal and other forces."

—A.H. Wilson, *CAHS Journal* 3.3 (Fall 1965), p. 68.

Here we see Alexander Eaglerock Combo-Wing G-CAIS (*The Silver Eagle*). It served B.C. Airways for flying lessons and revenue generation—charter flights and sightseeing rides. A three-seater, powered by a 90hp Curtiss OX-5 engine, it was a more serviceable aircraft than the Driggs Dart G-CAIR, a small and underpowered two-seater. Early in the summer of 1928, B.C. Airways purchased a second Alexander Eaglerock, registered as G-CATN.

The upper two photographs show different views of Eaglerock G-CAIS at Lansdowne Field. In the bottom picture, Hal Wilson (right, in flying helmet) has just given a sightseeing ride to John Satta Sr. of 2653 Dalhousie Street, who is gingerly climbing down from the wing after his first-ever flight. Only a handful of Victorians had ever flown before 1928, and a great many lined up to experience that thrill. The typical cost for a short ride was five dollars.

Registered as G-CATN on July 12, 1928 was B.C. Airways' second Alexander Eaglerock. Here members of the Victoria Motorcyle Club line up before being taken in groups of two on aerial sightseeing tours over Victoria. The photograph likely belongs to the second half of 1928.

Originally expected to start in April, and then in May, scheduled flights from Victoria to Vancouver and Seattle did not begin until late July. We should not be surprised that the starting date of passenger service could not be precisely foretold. The chosen aircraft was a very hot commodity, and the Stout Metal Airplane Division of the Ford Motor Company (Dearborn, Michigan) was no doubt having a hard time keeping up with orders. B.C. Airways' machine, only the twenty-sixth to come off the assembly line, was well worth waiting for. And there were many preparations to be made in Victoria.

Flight instructor Hal Wilson was grooming a pair of his student aviators to serve as "assistant pilots" in the cockpit of the new airplane (Wilson 1965: 68). The term "co-pilot" was not yet in general use—even the concept was unfamiliar—and the title "First Officer" did not appear in B.C. aviation until the second half of the 1930s.

With B.C. Airways from the start, these two young men had shown unusual promise. One of them acquired posthumous fame when he perished aboard the ill-starred Trimotor. Robert Lewis Carson, born May 7, 1908, lived with his father, T.M. Carson (a restaurateur), at 2741 Asquith Street, within walking distance of Lansdowne Field. We have that information in his own handwriting, on a Victoria High School registration card (1924–25, Division 6), still extant in the archives of that school. Young Lewis —not Louis, as his name is usually spelled—was known to his friends as Lew. By a macabre twist of fate, he earned his commercial pilot's licence (Canada No. 337) on August 24, 1928, the very day before the crash (Molson 1980: 61). There is no reliable photograph, it seems, other than this murky image, from the *Victoria Times* of August 27, 1928.

Lew Carson (1908–1928)

In 1965, Hal Wilson—B.C. Airways' pilot on Victoria–Vancouver flights—identified Lew Carson's counterpart on that run as W.A. Hughes. Hughes must have been flying on a temporary licence, like the American veteran pilot Herold Walker. This is presumably the same W.A. Hughes who earned commercial licence No. 356 on September 29, 1928, and worked later for Canadian Colonial Airways of Montreal (Molson 1980: 61).

There was a flurry of excitement in May 1928, when two would-be competitors arrived on the scene. One was Western Canada Airways of Winnipeg, which had bought Vancouver's Pacific Airways and was planning a Vancouver–Victoria flying boat service. From Seattle, Percy Barnes and Vern Gorst had also set their sights on Victoria with their Boeing *Zephyr*, on which they persuaded the mayors of Seattle and Victoria to fly.

On July 8, however, Ernest Eve could proudly release the first official schedule for a tri-city air service, with Victoria at the hub. Indeed, a Victoria delegation, including his brother Cecil, pilot Herold Walker, and mechanic Ted Cressy, was already in Detroit, learning how to operate the aircraft so as to bring it home. Behind the scenes, Victoria insurance executive Fred McGregor was working frantically with other local businessmen to guarantee an emergency bank loan for $70,000—an enormous sum in 1928 dollars—so that B.C. Airways would be able to take possession of the aircraft and fly it west across the U.S. to Seattle and then to Victoria.

City of Victoria Archives 98403-29-6823

En route westward from Dearborn, Michigan to Seattle, July 19, 1928.
L–R: Ted Cressy; Cecil Eve; William Strain, president of Northwest Air Services Inc., Renton, Washington—Herold Walker's former boss; Corrine Motherall (Cecil Eve's mother-in-law); Mrs. Ruth Walker, enjoying a magical honeymoon; and pilot Herold (Hal) Walker. In front is the dog P.T., Walker's wire-haired terrier mascot.

City of Victoria Archives 98403-29-6821

Left. The evening of July 19, 1928. L–R: Ted Cressy, Cecil Eve, Herold Walker. A serious man of few words, pilot Walker was known as "Silent Herold." With 3,000 hours of flying experience in a wide variety of aircraft, he was unquestionably a well-qualified and competent aviator.

Below. Reaching Seattle at 1:30 p.m. on Sunday, July 22, the plane landed at Bryn Mawr airport in Renton, at the southern tip of Lake Washington. In this picture the official party is gathering on the morning of July 23. Departure for Victoria was delayed two hours while Cressy repaired an engine valve.

Gordon Cameron Collection

WEATHER FORECAST

# Victoria Daily Times

TIMES TELEPHONES

VOL. 73 NO. 17 — VICTORIA, B.C., SATURDAY, JULY 21, 1928 — 36 PAGES — PRICE FIVE CENTS

## WORLD RACERS ON WAY TO RECORD

### CARS PLUNGE DOWN FROM OPEN BRIDGE

Twelve Train Passengers Injured, But no Deaths in Accident on Long Island

One Car Three-quarters Submerged; Firemen Enable All Aboard to Escape

New York, July 21.—A dozen persons were injured to-day when the first two cars of a Long Island Railroad electric train plunged through an open drawbridge over

ROUND-THE-WORLD FLIERS PREPARING FOR TAKEOFF FROM VICTORIA

### Mears and Collyer Leave Victoria in Early Hours On Final Lap of Flight

### LOG OF WORLD FLIGHT PLANE

Famous Globe Encircler Calls at City for Second Time in Fifteen Years in Course of Epochal Journey; Takes Off at 3.37 a.m. for New York via Spokane and Minneapolis; Plane Is Tuned Up and Refueled Here for Dash Across Continent; Story of Flight

Silvery grey light was just appearing in the eastern sky this morning as John Henry Mears and Charles B. D. Collyer took off from the B.C. Airways Aerodrome on Lansdowne

By now, only a little more than a year had passed since Charles Lindbergh's solo flight to Paris, but it seemed in the summer of 1928 as if every man, woman, and dog was engaged in daring competitive flights. Speed records were all the rage, and the ultimate challenge was to circle the globe.

The most famous globe-trotters of 1928 were the American team of John Henry Mears and Charles B.D. Collyer. Back in 1913, in true Jules Verne fashion, Mears had set a world record that lasted until 1926; now he was flying as a passenger with Captain Collyer. In just over three weeks, from June 29 to July 22, the two set a new speed mark in their flight from New York to New York, crossing the Atlantic and Pacific by steamer—fair game by rules of the day. Their aircraft, a single-engine Fairchild FC-2W folding-wing monoplane (NX-5501), named the *City of New York*, was destined to end its days in the Canadian north as CF-ACZ. By a splendid coincidence they passed through Victoria via Lansdowne Field on July 20–21, even as Herold Walker was flying B.C. Airways' new Ford Trimotor west from Dearborn. What is more, their triumphant arrival in New York was reported on the front page of the *Victoria Times* right below its account of the Trimotor's landing in Victoria.

The Mears-Collyer event put Victoria briefly in the world media spotlight, for the same reason that Eddie Hubbard had been able to make such a success of the Victoria–Seattle airmail service. The explanation is simple: the fastest ocean liners then crossing the Pacific made Victoria their first North American port of call.

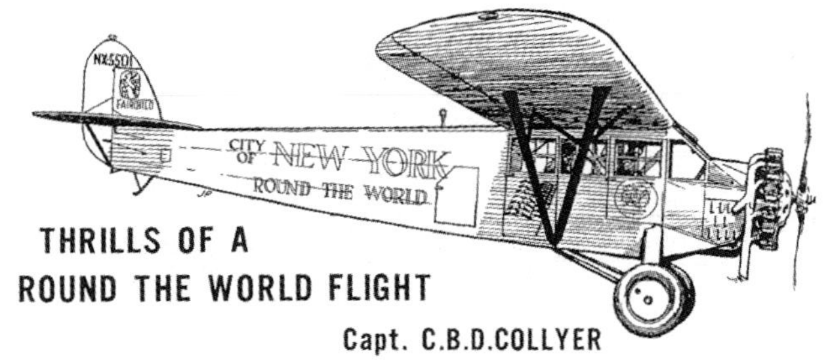


*CAHS Journal*, 8.3 (Fall 1970): 78–85

Accompanied by the inevitable canine mascot, a cute little Sealyham terrier named Tailwind, the adventurers arrived in Victoria on the evening of Friday, July 20 aboard Canadian Pacific's *Empress of Russia* from Yokohama—ten hours ahead of schedule. In a blaze of lights, newsreel cameras recorded the liner's arrival at William Head, where reporters crowded aboard the quarantined vessel to interview the two heroes even before they disembarked on the Outer Wharf at 10:10 p.m.

With wings still folded, the plane was towed behind a truck to Lansdowne Field, where work was hastily started to ready it for the final leg of the journey to New York. Wilbur Thomas, western representative of the Pratt & Whitney engine company, was on hand to service and tune the motor, with drums containing 212 gallons of gas to fuel the tanks. By 3:30 a.m. on the morning of July 21, the engine was warmed up, and all was ready. The two men climbed aboard the aircraft with little Tailwind, a flare-path of gasoline pots was set a-blazing, the Fairchild's motor roared, and they were airborne at 3:37 a.m. A large crowd of well-wishers saw the *City of New York* bank towards the east, where its dim shape soon vanished into the pre-dawn haze. It was apparently heard passing over Seattle's Boeing Field at 4:28 a.m., no doubt to the chagrin of local enthusiasts. The next stop to refuel was Spokane, then on to Minneapolis. On the day following their Victoria departure, the two men (and Tailwind) landed at New York, setting a new round-the-world record of 23 days, 15 hours and 21 minutes.

WEATHER FORECAST

# Victoria Daily Times

TIMES TELEPHONES

VOL. 73 NO. 18 — VICTORIA, B.C., MONDAY, JULY 23, 1928—20 PAGES — PRICE FIVE CENTS

# PLANE ARRIVES TO START NEW SERVICE

## ACCIDENTAL DEATH IS THE JURY FINDING

Inquest Held To-day Over Body of Douglas Peterson, Bank Teller

Found Dead in Room Over Bank Premises; Accounts in Order

Accidental death was the verdict brought in by the coroner's jury this morning investigating the death of Douglas Peterson, teller of the Canadian Bank of Commerce, corner of Oak Bay Avenue and Foul Bay Road, who was found dead in a room above the bank on Saturday morning with a bullet wound in his head.

The inquest was held at the B.C. Funeral Parlors with Coroner E. C. Hart in charge.

Sarah Short, janitress at the bank, said the teller arrived at the bank at 8.35 o'clock, which was exceptionally early for him. Proceeding to the teller's cage he hung up his coat and then went across to the corner store to see if the mail had arrived. Coming back he entered the cage and then went to the quarters above the bank. Shortly after witness declared she heard a noise. Going upstairs she saw Peterson lying stretched out on the floor in a bedroom with a bullet wound in his head.

## TRI-MOTORED FORD MONOPLANE OF TYPE ENTERING SERVICE BETWEEN SEATTLE, VICTORIA AND VANCOUVER

## MacLean to Consult Tolmie on Date Of Cabinet Retirment

## ARMED CITIZEN FOILS ROBBERS

Vancouver, July 23. Burglars were carrying a safe from a groceteria here early this morning when George Fox, who lives next door, shouted "Drop it." They did so when Fox backed up his order with a shotgun. The thieves escaped in a stolen automobile, which they later abandoned.

## WORLD TRAVEL RECORD IS SET

Mears and Collyer Circle the Globe in 23 Days 15 Hours 21 Minutes

## Big Ford Plane For B. C. Airways Gulf Service Here After Flight From Detroit

Pilot Herold Walker Brings Trimotored Monoplane to Earth at Aerodrome at 1 p.m., Completing Hop From Seattle After 2,700 Miles Flight; Mayor Pendray and Mayor Taylor, of Vancouver, Make First Trip With Other Passengers; Service to Vancouver and Seattle Will Start This Week

By Times Staff Representative

Flying history was made in Victoria to-day with the arrival of the new Ford passenger air liner for the B.C. Airways Limited gulf service. The huge trimotored plane was brought to a gentle landing at the Lansdowne Aerodrome at 1 p.m., with Pilot Herold Walker at the controls. A crowd of several thousand people was on hand to witness the finish of an epoch-making flight.

Carrying Mayor and Mrs. J. Carl Pendray and Mayor L. D. Taylor, of Vancouver, and several flying officials and other passengers, the big air liner flew from Seattle to-day, completing a flight of 2,700 miles from Dearborn, Michigan. It took off from Seattle at 11.52 o'clock, after being delayed by slight trouble with one of its three engines.

CARRIED PASSENGERS

The passengers on the inaugural trip

## BISHOP OF BUDAPEST IS NOW IN CITY

Count John Mikes Bound to Australia to Attend Eucharistic Congress

Victoria is One of Most Beautiful Cities He Has Ever Visited

Bishop Count John Mikes, of Budapest, Hungary, who is bound to Australia to attend the Eucharistic Congress at Sydney, is an interesting visitor to the city to-day, and will sail for the South Seas by the liner Niagara when she puts to sea on Wednesday night.

Bishop Mikes, with his secretary, Rev. Dr. John Zabo, reached Victoria yes-

In July 1928, aviation kept the Victoria newspapers busy running banner headlines. By Monday, July 23, Mears-Collyer have been eclipsed, relegated to a secondary story beneath the Trimotor landing at Lansdowne Airfield.

Above is a classic J.Howard A. Chapman photograph of the huge Trimotor's arrival at Lansdowne Field on the afternoon of Monday, July 23, 1928. The U.S. registration numbers 5810 can be seen beneath its left wing.

The photo to the right captures the drama and excitement of the Mears-Collyer departure from the same field only two days earlier, in the pre-dawn hours of Saturday, July 21, 1928.

Elwood White Collection

## Seattle–Victoria Passengers on July 23 Aboard Trimotor 5810

(L–R) Pilot Herold Walker (with mascot P.T.); Mayor L.D.Taylor of Vancouver; Mrs. Ramsay and Claude Ramsay, U.S. National Aeronautic Assn.; Ernest Eve; Gordon Cameron, with son Donald; mechanic Ted Cressy (partly hidden); Mrs. Cameron; Mrs. Pendray and Mayor J. Carl Pendray; Mrs. Ernest Eve (in front); Cecil Eve, who was already in Victoria; Mrs. Cecil Eve (in front); Frank Fretwell, U.S. National Aeronautic Association. Missing: *Times* reporter Leslie Fox.

Gordon A. Cameron Collection

Festive crowds had gathered at Lansdowne Field to greet the Trimotor on its delayed afternoon arrival. Loud cheers arose at 12:55 p.m. as the plane approached from the southeast, passed over the airfield, and circled above Mount Tolmie. The suspense grew when Herold Walker appeared to abort his first approach and then circled again to come in from the south. As the wheels finally touched down on the grass runway, some over-eager fans broke through the police cordon. Some fifty years later, air engineer Ted Cressy vividly recalled the panic he had felt in performing his duties as assistant pilot:

"We were two hours late so they were getting hot and happy, as it was a beautiful day. We had to land downwind; we couldn't land in-wind at the time because of Bowker Creek. So we came in a little downwind, and Hal [Walker] brought her down just over Bowker Creek and we started rolling toward [where] the crowd was, all around the hangar, all the dignitaries were standing there, and he yelled to me, 'Pull on the brakes.' So I cut the motors and of course the motors took quite some time to unwind, and I hauled on this brake, left and right. It was like a joystick brake. I could see that crowd coming nearer and nearer, faces were getting bigger and they were just standing there solid, and we stopped not more than 20 or 25 feet from the crowd, just as if it was beautifully arranged. But little did they know how many pints of sweat were running down the back of my spine. I was seeing those three propellers chewing up everybody and our hangar as well. That was an apprehensive moment. Moments like that made Mrs. Cressy's little boy wish he had gone into the grocery business!" (*The Magnificent Distances* (1980): 37)

### A PROFUSION OF LANSDOWNES

Victoria's Lansdowne Field was located on Lansdowne Road. This street name was a tribute to the Marquess of Lansdowne, Canada's popular Governor General from 1883 to 1888. Widely adopted across Canada, his name had been given also to a racetrack on Lulu Island and a nearby field used briefly for aviation. It can be quite confusing that B.C. had two Lansdowne airfields.

A close call, perhaps, but a triumphant arrival—followed by a luncheon and civic reception in the Dominion Hotel. At 3:25 the same afternoon, a different group (mainly B.C. Airways Directors) took off for Vancouver, where yet another throng was waiting at that city's Lansdowne Field, on Lulu Island. As he rushed from the plane at 4:05, eager to meet well-wishers, Vancouver Mayor Louis D. Taylor stepped directly into a still-spinning propeller and incurred a potentially fatal skull fracture. It was nearly a disaster, but miraculously he survived, and was soon back on the job.

Donald Cameron, 20 months, first "junior" ever to fly on an air service between the U.S.A. and Canada

Gordon A. Cameron Collection

## Excursion as G-CATX to Seattle for Dedication of Boeing Field

The date is Thursday, July 26,1928. New Canadian registration letters have been painted on the Ford Trimotor (now G-CATX), and all is in readiness for a 2 p.m. flight to Seattle, where this Victoria group will be guests of honour at the dedication of the new Boeing Field. It is the ideal moment for a formal photograph, a picture that will appear on the front page of the July 28 *Daily Colonist*, and will then be used in promotional advertisements.

(L–R) Alex T. Stewart (advertising manager of the *Colonist*), Cecil Eve (general manager of B.C. Airways), John A. Rithet (see below), J. Ernest Eve (president of B.C. Airways), Victoria Mayor J. Carl Pendray, A. Haliburton Wilson (assistant pilot; not the pilot, as the *Colonist* states), Mrs. Carl Pendray, pilot Herold Walker (whose mascot dog, P.T., lies at his feet), his new bride Ruth Walker, Victor R. Gravlin (*Colonist*), Charlie R. Walker, Ted Cressy (B.C. Airways' air engineer, in back row), Clifford E. Whitaker, William B. Wetmore, Francis E. Winslow, and J.E. Todd.

The five men on the far right are revenue passengers, who have business in Seattle. Not shown is John Shaw of the *Times*; Ted Cressy is to stay in Victoria. Two young neighbourhood boys have furtively joined the group and peek through from the back row; one of them may be Elwood White. Mr. Valdemar F. Bendrodt, traffic manager of B.C. Airways, made the return trip—replacing Mr. Rithet, whose wife was ill in a Seattle hospital. John A. Rithet (1878–1942) was the aviation-boosting son of R.P. Rithet, of Rithet's Outer Wharf and Rithet's Bog.

Gordon A. Cameron Collection

The huge aircraft had a powerful attraction for local youngsters. Has this pair been swimming at Willows?

Before the plane began regular service on July 30, there were courtesy flights for prominent Victorians. This may be one, as the man in the centre, facing left, is B.C. Airways lawyer Gordon Cameron, a veteran of the old Aerial League of Canada (Victoria Branch).

**BRITISH COLUMBIA AIRWAYS LTD.**

Agent's Stub (to be detached when sold)

A

ONE WAY

No. 3106

From ..............................

To ..............................

Sold by ..............................

Date ..............................

**BRITISH COLUMBIA AIRWAYS**
LIMITED

**NON-TRANSFERABLE**
**ONE WAY**

Sold subject to regulations printed on back

When officially stamped on back, good for one continous passage as indicated on coupon attached.

In selling this ticket and in checking baggage hereon, the British Columbia Airways acts only as agent and is not responsible beyond its own lines.

.............................................. 192.......

Reservations for passage on this ticket is made for..............................

..............................
(Signature of passenger)

..............................
(Agent—Witness)
CECIL H. EVE,
Manager.

A Void if detached

No. 3106

**BRITISH COLUMBIA AIRWAYS LTD.**

Good for one passage by air

From ..............................

To ..............................

Price.................. Sold by..................

A

No. 3106

**BRITISH COLUMBIA AIRWAYS LTD.**

Passenger Identification Check
**Not Good For Passage**

From ..............................

To ..............................

Price.................. Sold by..................

# BRITISH COLUMBIA AIRWAYS
LIMITED

BCA

VICTORIA—VANCOUVER—SEATTLE
DAILY
(Three-motored, all-metal, 14 passenger Ford Air Liner)

| LEAVE | | ARRIVE | |
|---|---|---|---|
| Victoria, | 8.30 a.m. | Vancouver, | 9.15 a.m. |
| Vancouver, | 9.30 a.m. | Victoria, | 10.15 a.m. |
| *Victoria, | 10.30 a.m. | Seattle, | 11.30 a.m. |
| *Seattle, | 3.00 p.m. | Victoria, | 4.00 p.m. |
| Victoria, | 4.15 p.m. | Vancouver, | 5.00 p.m. |
| Vancouver, | 5.15 p.m. | Victoria, | 6.00 p.m. |
| *Vancouver, | 9.30 a.m. | Seattle, | 11.30 a.m. |
| *Seattle, | 3.00 p.m. | Vancouver, | 5.00 p.m. |

* No service to Seattle on Sunday.

RATES

*Between* VICTORIA and VANCOUVER—
One way, $11—Return, $20.

*Between* VICTORIA and SEATTLE—
One way, $12—Return, $22.

*Between* SEATTLE and VANCOUVER (*via* Victoria)
One way, $18—Return, $35.

TICKETS, INFORMATION, RESERVATIONS

VICTORIA: (Head Office) Eve Bros., 900 Fort St.
Phones 2552 and 1969.

VANCOUVER: B.C. Motor Transport Depot,
Seymour and Dunsmuir Streets.
Phone Seymour 4000.

SEATTLE: Motor Bus Depot, 1918 3rd Avenue
Phone Elliott 3565.

*Use the Air Mail and Air Express—They Save Time*

The original schedule (*Times* 1928.07.28: 7) called for four daily departures from Victoria: two to Seattle (8:30 a.m. and 1:45 p.m.) and two to Vancouver (11:00 a.m. and 4:15 p.m.). That was soon found to be impractical and uneconomical; the revised schedule above was in effect on the day of the crash. Our thanks to Donald G. Cameron for letting us reproduce the original B.C. Airways ticket (left), from the collection of his father, Gordon A.Cameron.

Elwood White Photo

At age 10, Elwood White used his Brownie camera to snap a rare picture of the Trimotor in flight over Lansdowne Field (right), and later collected the special newspaper sticker (top) from the August 11,1928 in-flight issue of the *Vancouver Sun*.

VANCOUVER ····· 45 MIN. ····· VICTORIA ····· 1 HR. ····· SEATTLE

*By*

# AIR!

*To and From*

VICTORIA

SEATTLE *and* VANCOUVER

*"The Wings of Modern Business!"*

*Safe and Sane*
*Ford Three-Motored All-Metal*
*14-Passenger Cabin Air Liners*
*of the*

BRITISH COLUMBIA
AIRWAYS LTD.

| VANCOUVER | (HEAD OFFICE) VICTORIA | SEATTLE |
| --- | --- | --- |
| 1501 Bekins Bldg. | 900 Fort Street | 1918 3rd Avenue |
| Phone Seymour 4684 | Phones 2552 and 1969 | Phone ELliot 3565 |


*By*

AIR!

*To and From*

VICTORIA

SEATTLE *and* VANCOUVER

*Safe and Sane*
*Ford Three-Motored All-Metal*
*14-Passenger Cabin Air Liners*
*of the*

BRITISH COLUMBIA
AIRWAYS LTD.

VANCOUVER 1501 Bekins Bldg. Phone Seymour 4684
(HEAD OFFICE) VICTORIA 900 Fort Street Phones 2552 and 1969
SEATTLE 1918 3rd Avenue Phone ELliot 3565

This special British Columbia Airways Limited 5¢ blue airmail stamp was meant to supplement a regular 2¢ Canadian stamp for a total air-mail letter cost of 7¢. Canadian postal authorities granted a Victoria–Vancouver air-mail contract and approved this rouletted stamp, designed by Cyril Connerton of Victoria. About 100,000 copies were printed, but relatively few were used after the first day of issue on Friday, August 3, when approximately 500 pieces of mail were handled. The topic was researched and written up by Vancouver philatelist H.L. Banner, in an undated and perhaps unpublished paper entitled "British Columbia Airways Ltd." Dated covers have become very valuable, but mint stamps are relatively common.

Again, we express our thanks to Donald G. Cameron.

Reproduced above, about half size, is part of a glossy promotional brochure published by British Columbia Airways in the summer of 1928. Text and graphics are printed on both sides of one 17 x 11" sheet of paper, trimmed and folded into an eight-page booklet. Our photo on page 45 of the wicker seats in the plane's cabin is taken from this brochure. The 14-passenger Ford Trimotor is praised to the skies, with a repetitive emphasis on its safety.

During that midsummer of 1928, the thrills came in rapid succession. On Monday, July 30, B.C. Airways began its scheduled service (soon to change) with a flight to Seattle at 8:15 a.m., an event reported in the afternoon *Times*: Giant Plane Starts Hops to Mainland. But a larger headline in the same paper read VANCOUVER BOY IS OLYMPIC WINNER. The dazzling Ford Trimotor had been upstaged by the more dazzling Vancouver sprinter Percy Williams, "the world's fastest man," who won both the 100 and 200 metre races in Amsterdam.

Within the next year, Canada's Olympic double gold medalist flew over to Victoria on B.C. Airways' second Eaglerock, G-CATN. This picture shows a smiling Percy Williams (left) at Lansdowne Field, with pilot Hal Wilson seated centre and Vancouver air engineer William T. Bolton standing right. Between Bill Bolton and the majestic automobiles, we have a good view of the Shell Oil gasoline pump.

The all-metal monoplane was powered by three 220hp Wright J-5 engines. This Ford model (4-AT-B) normally accommodated two pilots and twelve passengers, but the Victoria aircraft was equipped with a folding seat that raised the passenger limit to fourteen. The wing span was 74 feet, almost three times that of Driggs Dart G-CAIR.

For twenty-six days, from Monday, July 30 to Friday, August 24, the splendid Ford aircraft performed magnificently, carrying 1,292 passengers on some 150 scheduled flights, with Victoria and Vancouver passengers comprising 75% of the traffic (Wilson 1965: 68). It was on the last Saturday in August 1928 that B.C. Airways' blissful world collapsed. Here is the first story in the *Seattle Post-Intelligencer* of a presumed disaster the day before; reporter R.B. Bermann sets the stage in an almost Gothic prologue:

> Into the thick, dank fog which cloaked the Strait of Juan de Fuca like a funeral pall the great Ford monoplane of the British Columbia Airways soared yesterday morning with five men and a woman aboard, bound from Victoria to Seattle.
>
> And like some fabulous, gargantuan monster, the fog reached out its long chill arms, enfolding plane, crew and passengers.
>
> From that dread embrace the plane had not emerged more than twelve hours later.
>
> She was seen struggling with the fog at about 11 a.m., half an hour after she had left Victoria, when she passed over the Dungeness lighthouse, just across the strait.
>
> (*Seattle Post-Intelligencer* 1928.08.26: 1)

Only one factual error slips into this account: there were *six* men and a woman aboard, although that other man—the fifth passenger—was not on the original manifest. In due course, it was confirmed that the giant aircraft had plunged into the sea not far from Dungeness, and the seven victims could all be correctly named:

Herold Walker, pilot, Seattle and Victoria
Robert Lewis Carson, assistant pilot, Victoria
Dr. Donald Booth Holden, Victoria
Mr. and Mrs. Alexander MacCallum Scott, Glasgow
Thomas E. Lake, Shanghai
Dr. Floyd D. Soverel, East Orange, New Jersey

There were several reasons why the disaster became a huge international news event—given almost as much initial coverage in the *Times of London* as in the *Victoria Times* or *Colonist*. To begin with, this Ford Trimotor was the largest, most modern commercial aircraft in the world. Even though it was carrying only a partial load of passengers, the accident could be accurately described by a U.S. government inspector as "by far the worst catastrophe in the history of commercial aviation in America" (*Seattle Post-Intelligencer* 1928.08.30: 3).

Second, there were powerful elements of mystery and suspense. Four days passed before the crash site could be even roughly located; it was never exactly pinpointed, because debris from the wreckage kept shifting. In the meantime, a massive aerial search mobilized almost every aircraft in the Pacific Northwest. A controversy arose over the conflicting evidence of "eyewitnesses"; some placed the crash southeast of Port Townsend.

Third, the relatively small passenger list involved some unusually interesting names and stories. Alexander Scott, a British barrister, was formerly a member of the House of Commons, a parliamentary secretary to Winston Churchill, and the Liberal Whip for Scotland; he and his wife had been spending some time at the Empress Hotel. Thomas Lake, who had just arrived in Victoria aboard the *Empress of Canada*, was the Far Eastern manager of the Union Oil Company of California; his father, Simon Lake, was a famous inventor of submarines.

Only on the Tuesday (day 4) was there firm evidence that a fifth passenger had been aboard. While flying from Vancouver to Victoria on the Saturday morning, Dr. Soverel had made a spur-of-the-moment decision to stay aboard the aircraft and continue to Seattle. His seat-mate, C. Crawford of Vancouver, now reported this fact.

"When we reached Victoria," said Mr. Crawford, "he said to me: 'This plane goes on to Seattle, doesn't it? I'll stay on board if it does.'

"He did not disembark." (*Times* 1928.08.28: 1)

The very next day (Wednesday, August 29), Dr. Soverel's body was one of the first two to be found, on the shore of Discovery Bay, to the west of Port Townsend. Also recovered was the body of Mrs. Jessie Scott (herself a very prominent Glaswegian), with her husband's leather briefcase and diary, and a telltale wicker chair.

Seattle Mayor Frank Edwards reported his own narrow escape from death. A last-minute Saturday commitment in Victoria had caused him to cancel his return reservation on the Ford, aboard which he had flown up from Seattle.

Human interest stories abounded. The Victoria papers revealed, for the first time, that Ruth and Herold Walker became married in Dearborn, Michigan; the 2,700-mile flight westward in mid-July had been their honeymoon. Her parents and his mother lived in Seattle.

Victoria grieved for Lewis Carson, a fine, ambitious young man who was a licensed pilot for one tragic day.

The great shocker in Victoria was the death of Dr. Donald Holden, probably the best known and most popular physician in the city. For many years his office at 851 Fort Street, near Quadra, had also been his home, though he and his wife Madge had recently built a new house on the View Royal waterfront. This was his first-

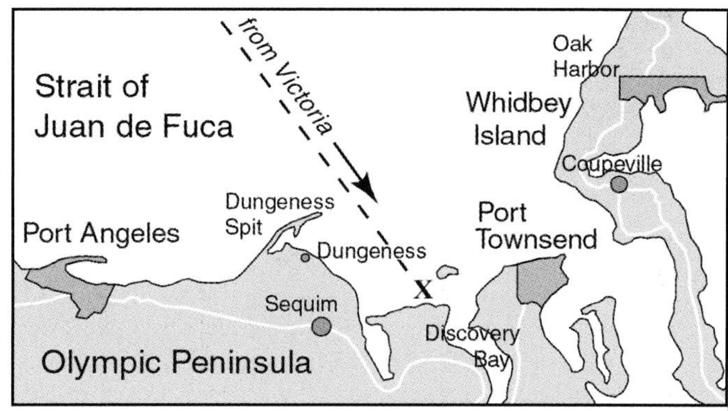


ever airplane ride: he was off to Seattle to join Madge, who was visiting her son Alex Holden, forest patrol pilot for the Washington Aeronautical Corporation of Tacoma.

Alex was Donald's stepson, but also his nephew, as Madge had been previously married to Donald's older brother Frederick. Her eldest son Clement was born in Montreal, but Alex and his two sisters were native Victorians. Madge and all her family had lived since the 1890s with the children's bachelor uncle, a surrogate father to Alex long before he and Madge got married in 1920. Born on November 11, 1895, Alex was now 32.

Though he had grown up in Victoria and worked in town for some years after wartime service with the Royal Flying Corps, Alex Holden was not a familiar name to most Victorians in 1928. He was, however, one of the most conspicuous pilots in Seattle, a hotbed of aviation. The Seattle papers made much of his heroic and poignant leadership in the aerial search for the missing Canadian aircraft. Like the Victoria–Seattle airmail pilot Percy Barnes, he risked his life by flying in dreadful conditions of fog, smoke, and overcast, aloft from dawn to dusk, from Saturday afternoon until at least Tuesday night.

Courtesy Gloria Holden Simpson

Dr. Donald Booth Holden

**LOCAL MEDICAL MAN ON PLANE**

DR. DONALD B. HOLDEN, B.A., M.D., C.M., McGill, was senior member of the Victoria medical profession, having opened practice here in 1891, following brief service in Golden, B.C. He was sixty-two years of age, and was born at Belleville, Ontario, his father being a partner in the well-known firm of Ames, Holden & Company of Montreal (*Times* 1928.08.27: 1).

Photo by Lloyd Jarman

Alexander Bancroft Holden

How well did B.C. Airways fare in the wake of the August 25 disaster? During that horrible last week of August, its behaviour was exemplary—unstinting efforts in the search for the missing plane, with compassion and support for the victims' families. Immediately after the 2 p.m. phone call from Renton reporting that G-CATX had not arrived, Ernie Eve and Hal Wilson took to the air and headed southeast across the Strait to investigate. Later that afternoon, in separate planes, Wilson and Cressy joined U.S. aircraft in mounting an intensive search—all to no avail, of course. Though B.C. Airways tried to carry on, the tragic event foretold the end of a brave and promising venture in Canadian civil aviation.

You can still get an argument about what really happened on the fateful Saturday. The enquiry revealed that, although the Seattle weather bureau had reported fog and low ceiling, it had not recommended cancellation of the flight. There are some who have said that Ernie Eve was always too quick to discount safety concerns, but there is no evidence that he disregarded advice on this occasion. Years later, Ted Cressy said that "Hal" Walker had been goaded into taking off by the imperious former M.P., Alexander MacCallum Scott:

> "The story of that accident was that we got a telegram giving us the weather, and it was not very good. Walker had decided not to go when a passenger by the name of Scott said to him that he had been flying with Imperial Airways when you couldn't see your hand in front of your face. He said, 'What's the matter with you? Can't you people fly an airplane?' and Walker said, 'Get in.' And they took off."
>
> —*The Magnificent Distances* (1980): 38

No one knew the flight path better that Herold Walker, a pilot with impeccable credentials. He would have been familiar with the late-summer smoke hazard from forest fires, which compounded the problem of low-lying fog. (His successor on the airmail run, Percy Barnes, had twice landed his flying boat earlier that morning, before turning back to Seattle.) Hal Wilson and other experts suggested that Walker, as a flying-boat veteran, may have been in the habit of keeping too close to the water, with the result that he caught a wing-tip when banking in order to return to Lansdowne Field. Years later Wilson casually revealed that neither he nor Walker had ever learned to use the Ford's bank-and-turn indicator (Wilson 1965: 68).

*Seattle Post-Intelligencer* 1928.08.24: 12

By coincidence, this picture of Alex Holden, seated (left) as pilot of his Forest Patrol Stearman, graced the Seattle P-I's Aviation page the day before the Trimotor crash. On August 10, the P-I had reported Alex's victory, flying a Swallow, in the Seattle-Centralia Derby; Harold Bromley, another native Victorian, finished second; Anscel Eckmann, a Seattle pilot well-known in Victoria, was third.

Whatever happened that day, B.C. Airways was completely exonerated from blame in a September 1928 enquiry. That commission recommended the use of seaplanes or amphibians on routes between Victoria and the mainland, adding fuel to a continuing controversy in the autumn. Two veterans of the Royal Naval Air Service, Hal Wilson and Sydney Pickles (next page), spoke out as champions of land planes and seaplanes, respectively. Wilson may have been right, but it would be fifteen years before another land-based commercial air service linked Vancouver Island and the mainland of North America. Now that their confidence was shaken, it is doubtful that Victorians would have used such a service, had it existed.

On August 30, 1928, what was almost certainly the main wreckage was hooked at seventeen fathoms by a patrol boat near the entrance to Discovery Bay, but it slipped away from the grapple. The gruesome discovery of additional bodies continued until November, when all had been found and identified but the cockpit crew—which included the almost forgotten little terrier mascot. Dr. Holden's body was discovered in Port Angeles harbour on October 1, a week after his memorial service in Christ Church Cathedral. His appointment book and a B.C. Airways brochure were on his person.

Almost ten years later, on January 20, 1938, fishermen between Port Townsend and Dungeness recovered several pieces of barnacled wreckage: metal struts, an unopened parachute, an unignited flare, and a wheel with a badly burned tire. Although the report of fire damage is mystifying, this debris was thought to have come from the B.C. Airways crash (*Colonist* 1938.01.21: 3).

In mid-February 1928, a new Victoria Aero Club had been formed, taking advantage of a Government program to encourage Canadian civil aviation at the grassroots level by providing de Havilland Moth biplanes, under specific conditions, as teaching and practice machines. This plan was not unlike the scheme that created the Aerial League of Canada nine years before. Because the terms called for a nucleus of experienced flyers, Ernie Eve stepped into the background and let Great War aviators run this show.

Club members included familiar names from the old Aerial League of Canada, like Gordon Cameron and Jimmy Gray, but its officers had not been locally prominent as flyers, and most had no connection with B.C. Airways. The Club was pleased, however, to share the use of Lansdowne Field. Members who guaranteed the $4,000 bond required for the government "loan" of the D.H. Moth were A.W. (Nick) Carter, L. de S. Duke, D.W. Hanbury, R.H.B. Ker, H.M. McGiverin, W.L. Stewart MacLeod, Sydney Pickles, and Norman Yarrow.

President Alfred Williams Carter, M.B.E., D.S.C., had been a Captain in the R.F.C.; he held Canadian commercial pilot's licence No. 98, earned on January 31, 1920. Hon. Secretary, Australian-born Sydney Pickles, acquired his commercial licence (No. 266) on March 20, 1928, when the Aero Club was beginning. He shared Carter's conviction that Victoria could and should be the major Canadian airport on the Pacific Coast, and promoted that cause vigorously. Sydney Pickles (see photos next page) is well remembered locally as a Reeve of Saanich.

The Club's small hangar appears in many pictures. In the photo at top, the B.C. Airways hangar is No. 1, and the Aero Club is No. 2. It was built by Club volunteers, mainly with materials donated by local business firms.

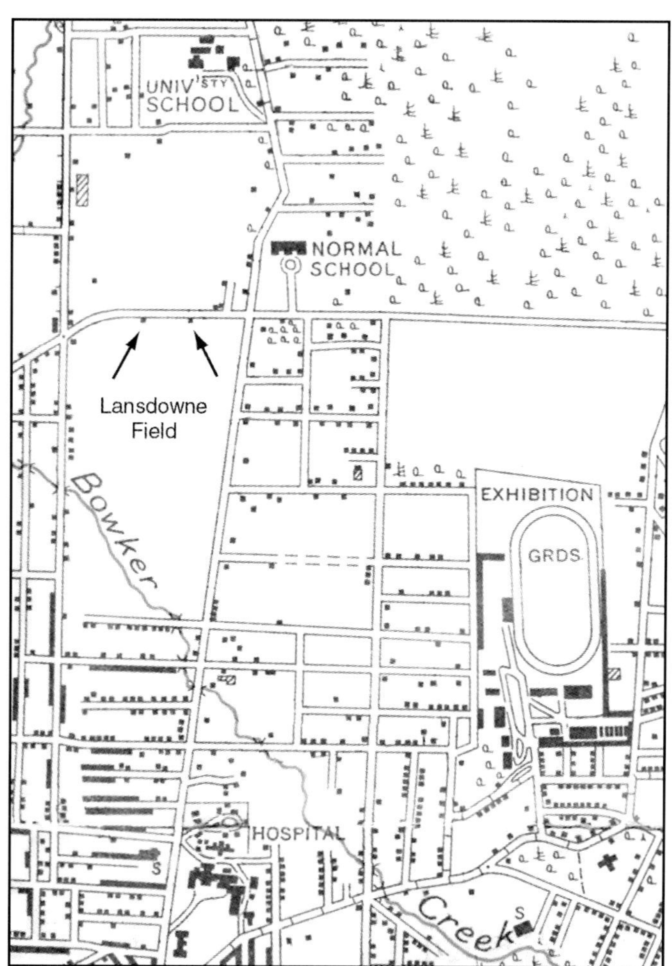


Both hangars appear on this edited 1934 topographical map ("Lansdowne Field" and arrows are added). There were few houses in the Saanich Panhandle between Royal Jubilee Hospital and University School—and fewer still in Oak Bay directly north of Willows Fairgrounds.

The Victoria Aero Club received its D.H. 60X Moth, registered as G-CAKA, at the beginning of July 1928; its arrival had been delayed for some months by the problem of raising the required bond. On July 2, after the plane had been passed as airworthy by Mr. Carter (a licensed air engineer as well as a pilot), it was checked out by Mr. Pickles. Three test flights were made, including many acrobatics.

Before long the Club had about a hundred members, with forty-two wishing to undertake flying instruction. Its first season of operation was successful, apparently not affected by B.C. Airways' problems after the August crash. Club members enjoyed a cooperative relationship with B.C. Airways employees like Wilson and Cressy.

President A.W. (Nick) Carter was at the controls of the Aero Club's D.H. Moth (G-CAKA) when it made its last flight. After only eight months of service, it came to a sudden end when it crashed on March 10, 1929 near the corner of Cedar Hill Road and Cedar Hill X Road. Carter was seriously injured, while his student pilot F.C. Aitken suffered shock and bruises. It was an intelligent mistake: forced to make an emergency landing, he chose an apparently ideal field, unable to detect from the air that it had been freshly ploughed. Goodbye, Aero Club!

Australian-born Sydney Pickles appears in both photos, taken when he was a Flight Lieutenant in the R.N.A.S.

On Friday, November 23, 1928, tragedy had again struck Lansdowne—the second disaster within three months. G. Arthur Raybone, a former Victorian living in Tacoma, and Valdemar Bendrodt, a well-known local man employed by B.C. Airways as Traffic Manager, spun out of control and crashed violently onto Lansdowne Field while flying in Raybone's combo-wing Eaglerock.

The two men had made a leisurely flight over the city and out to Cadboro Bay. Returning to the airfield they executed a few aerobatics after spotting the Aero Club's Moth being flown in that vicinity by Hal Wilson. Appearing to want a game of follow-the-leader, Raybone pulled up in a tight climbing turn as they were passing over the field. The machine stalled and dived into the ground, seriously injuring Bendrodt and killing Arthur Raybone. Both men had been R.F.C. pilots in the war. Arthur's father Enoch, a recent widower, still lived in Victoria; his son was buried in Ross Bay Cemetery.

Once again, Victorians read banner headlines lamenting an aerial calamity. We should note, however, that this was only the second flyer ever killed in a local accident. The other had been Johnny Bryant way back in 1913—then Canada's first-ever aviation fatality.

Another less horrific crash had occurred the previous month in Vancouver. In October 1928, the Driggs-Dart G-CAIR, christened *Miss Nan* only in January, was damaged beyond repair almost immediately after being sold and delivered to the Sprott-Shaw School of Aviation. The pilot survived, but the plane was a write-off.

## THE TACOMA CONNECTION

We have met three ex-R.F.C. pilots from B.C. who headed south to Tacoma during the 1920s in order to seek fame and fortune: Alex Holden and Harold Bromley (both born and raised in Victoria), and now Arthur Raybone.

Harold Bromley became world famous in July 1929 when he attempted to fly across the Pacific Ocean—Tacoma to Tokyo. As a crowd of 25,000 gathered on July 29 to watch him depart in a Lockheed Explorer named *City of Tacoma* (NR-856H), his vision was temporarily obscured at takeoff. His overloaded aircraft nose-dived beside the runway, turning the flight into a fiasco. For more on Bromley and his highly publicized failure in 1930 to cross the Pacific from Japan to the U.S.A., in collaboration with Australian navigator Harold Gatty, see our final chapter, page 199.

If asked to single out one Victoria flyer for special mention, we could find no better candidate than F. Maurice McGregor (1911–1995), who made a truly distinguished contribution to Canadian civil aviation. Flying instructor, stunt pilot, bush pilot, airline captain with both Canadian Airways and TCA, senior executive with TCA and Canadian Pacific, advisor to the Government of Pakistan, airline owner, entrepreneur—he compiled an astonishing list of "firsts," many associated with his native city of Victoria. In the course of this book we shall watch his career unfold.

Seven years younger than Maurice, Elwood White knew him from the early days at Lansdowne and Gordon Head; in later years they swapped stories and photographs. Thanks to Joyce McGregor, we have been granted access to Maurice's fascinating log books and his extensive collection of personal memorabilia.

Maurice had his first lessons in Eaglerock G-CATN, under Hal Wilson, from December 24, 1928 to February 13, 1929. The wintry scene above belongs to that period, when he was 17. On March 16, 1929, he first flew solo in Eaglerock G-CAIS.

Reproduced at right is Private Air Pilot's Certificate No. 401, earned by F. Maurice McGregor on October 21, 1929, at the age of 18. He was not allowed to hold a commercial licence until his nineteenth birthday in 1930.

When Hal Wilson and Ted Cressy moved to Vancouver in the summer of 1929, there was an almost total hiatus in Victoria aviation, forcing Maurice McGregor to complete his basic flying training in Vancouver. However, we shall see in the next chapter that he returned home in the early 1930s to make his presence strongly felt both at Lansdowne Field and the new Gordon Head Airport.

Maurice McGregor Collection

A. B. 16.-1-20-1,000

CANADA

No. 401

AIR REGULATIONS, 1920

File No. 1020-M-234

PRIVATE AIR PILOT'S CERTIFICATE (FLYING MACHINES)

This certifies that Fredwick Maurice McGregor,
whose address is 2126 Granite St,
Victoria, B.C.
and whose photograph is attached, is authorized by the Air Board to act as unpaid Pilot of Flying Machines not used for commercial or state purposes.

This certificate is subject to the conditions printed below and to cancellation at any time for cause.

Dated this 21st day of October, 1929.

For the Air Board.

~~Supt. Certificate Branch~~
Controller of Civil Aviation

Signature F. M. McGregor.

CONDITIONS

1. The holder of this Certificate shall not take up a passenger in any machine unless he has completed at least ten hours' solo flying. He shall not take up a passenger in any machine until he has, either alone or accompanied only by an instructor, flown a machine of that type for at least two hours and has flown a machine within six months for at least one hour.

2. The holder must pass a satisfactory medical examination and be certified as fit to fly within twelve months from the date of this certificate and within every twelve months thereafter. He must also, before flying after any serious accident or illness, pass a like examination and obtain a like certificate. The examination is to be made by a medical officer approved by the Air Board and the result thereof endorsed hereon.

Courtesy Joyce McGregor

## A Lansdowne Field Photograph Album

As it became known in 1928 that Victoria now had a reliable airfield, visiting planes began to arrive quite regularly from the western United States, where aircraft ownership of every kind was more widespread than in B.C. At the top of the page, three U.S. biplanes stand facing Lansdowne Road. The nearest is an Alexander Eaglerock (N-4753) and its right-hand neighbour is a U.S. Forest Patrol Stearman C-3B (C-6154).

In the second picture, a Forest Patrol Stearman stands with engine running before the B.C. Airways hangar. This is the same kind of aircraft flown by Alex Holden, Forest Patrol pilot for the Washington Aircraft Corporation of Tacoma (see page 64).

The third picture (left) shows a de Havilland 9 (U.S. Reg. 915) flown by the Richfield Oil Company. It had neon tubing under the lower wings for the purpose of aerial advertising.

Here in front of the B.C. Airways hangar is a Lockheed Vega 1, U.S. registration NC-199E. Named *Elizabeth Lind*, it was owned by George Westinghouse of Bainbridge Island, elder brother of Aubrey Westinghouse, later the founder of Victoria's Westinghouse Airways (see Chapter 19).

This aircraft was converted in 1931 to a Vega 2-D float plane, as the 220hp Wright J-5 engine was replaced by a 300hp P&W Wasp.

## Visiting Aircraft from the United States

The aircraft seen above at Lansdowne Field in 1930 or 1931 is a Fleet 7 (NC-737V), owned and flown by American writer Ronald de Levington Kirkbride. Several of his novels were turned into popular Hollywood movies.

Second from the top is a Ryan B-1 Brougham, the same type of aircraft as Lindbergh's *Spirit of St. Louis*. This plane, named *Queen of Cowlitz*, belonged to a Mr. St. Johns, who owned an automobile agency in Chehalis, Washington. His pilot at the time was Ray Howard, whose name can be read on the cabin door.

The familiar Provincial Normal School building is partly visibile below the upper wing of this U.S. biplane. It is a Buhl CA-5A Airsedan, U.S. registration 5311. This model, manufactured in 1928, was the deluxe version of a very fashionable aircraft.

This visiting Alexander Eaglerock A-2, manufactured in Colorado Springs, has just been registered in Canada as G-CAUZ *Northern Light*. The date is September 24, 1928. It is en route from Vancouver to Whitehorse to begin service on skis with Yukon Airways & Exploration Ltd. Clyde Wann, operating manager of the Yukon company, and pilot J.M. Patterson have given Ernie and Cecil Eve a "trial spin" *(Times* 1928.09.25: 13).

# 9. Lansdowne or Gordon Head? Two Lost Causes ▪ 1929–33

From 1927 through 1929, the development and administration of Lansdowne Field was a private enterprise, initiated by the Eve brothers and managed by British Columbia Airways, Ltd, in accordance with the rules and regulations of the Dominion Government's Department of Civil Aviation. Civic politicians from Greater Victoria showed up on major ceremonial occasions to express their approval and support, but their attitude towards local land-based aviation was basically one of laissez-faire. Because there were civic movers and shakers among the directors and backers of B.C. Airways, political influence and advice could perhaps be dispensed informally, but there was no municipal control. This situation would change dramatically during the years 1930 to 1933, when the City of Victoria became directly involved in airport management, taking charge of not just one, but two airfields—both in Saanich. On the momentous occasion of June 29, 1931, the future of air travel was submitted to a vote of Victoria ratepayers.

The progress of this controversial subject can be followed in the minutes from 1928 to 1933 of the City of Victoria's Industrial Committee, renamed on June 17, 1932 as the Business and Trade Development Committee. Aviation was not the only topic on its agenda, but aviation often dominated discussion. At times there were joint meetings with representatives of the Victoria Chamber of Commerce, a body with strong views about air travel. In the early stage (1928–30), the main focus was on the hot topic of an air harbour for seaplanes and flying boats, a subject that we address in Chapter 11. On September 30, 1929, there was keen interest shown also in an airport for land machines. Herbert Anscomb had succeeded Carl Pendray as Mayor of Victoria; the Chairman of his Industrial Committee was Alderman William T. Straith, a prominent lawyer who now became deeply engaged in civil aviation. (Two decades later, both Herbert Anscomb and William Straith would be cabinet ministers in the B.C. Coalition Government of Byron "Boss" Johnson.)

Within a one-week period (meetings of November 12 and 18, 1929), Bill Straith and his Industrial Committee negotiated and recommended to City Council a transfer of the lease for Lansdowne Field from B.C. Airways to the City of Victoria. The property had been leased from Mrs. Elizabeth Scott for a five-year period, but the Eve brothers were obviously eager to bail out. The revised terms called for a rental fee of $650 to October 1, 1930, and $700 per year for the next two years. Until October 1, 1930 the City would have an option to purchase the land at $1,000 per acre, or $77,000 in total. City Council swiftly approved these terms and assumed control.

In the months ahead, this Committee would deal with requests to herd cows on Lansdowne Field and cultivate portions of the land (no to the first, but a qualifed yes to the second—for a $50 contribution to the annual rental fee). More substantial were requests to operate aviation schools, in particular an application from the Sprott-Shaw School of Aviation, now thriving in Vancouver. After consultation with Squadron Leader Earl MacLeod (RCAF Jericho Bay), the City's current technical advisor, that type of activity was deemed no longer appropriate.

By August 4, 1930, the future of Lansdowne was looking much brighter. City Council approved a revised option to purchase, signed that evening with Mrs. Scott's lawyer, at the much lower price of $51,250 (with a new deadline of July 1, 1931). What is more, Major-General MacBrien was offering assurances that there would be no licensing difficulty (*Colonist* 1930.08.05: 1, 6).

Although the focus of attention would shift to the new Gordon Head Airport, Lansdowne continued to be used unofficially for years. In July 1935 Gordon Ballentine landed this D.H. Dragon (CF-AVD) at Lansdowne.

Even so, a notice was placed on Lansdowne Field in September 1930: "that commercial flying is prohibited and the Field is for 'emergency' landing purposes only." Concerns about safety led soon to the appointment of Mr. A.R.C. Morton, City Engineer's Department, as Manager of Lansdowne Field.

September 18, 1930: as if out of the blue, the Industrial Committee was discussing a new airport at Gordon Head: "The Chairman reported progress on suggested site for new airport in the Finnerty Farm District." By February 6, 1931, by quiet negotiation, Straith secured options on the needed Gordon Head properties. Lansdowne Field seemed now forgotten.

During the first half of 1931, events moved swiftly for the proposed new Airport. In February, even as B.C. Airways was trying vainly to resurrect itself at Lansdowne (see next page), Mayor Anscomb had a positive reception from Saanich Reeve Crouch in seeking a remission of taxes on the desired land at Gordon Head. The same month, R. Carter Guest, Dominion Inspector of Air Regulations, decreed that Gordon Head's Finnerty Farm was better than other alternatives to Lansdowne, which included Macaulay Plains, Esquimalt Lagoon, and Portage Inlet. On May 27, the Industrial Committee reviewed an estimate of construction costs totalling $234,000 (plus cost of land and financing).

Following Council approval, an Airport By-Law vote (for $240,000) was scheduled for Monday, June 29, 1931, and was promoted by large newspaper advertisements, promising—perhaps unwisely—that the total cost (unstated) would be spread over the next thirty years.

**Vote For the Airport By-law Monday June 29**

Perhaps no Victoria money by-law has ever gone down to a more lopsided defeat: 3,247 to 332. This was June 1931, the depths of the Great Depression, and City ratepayers were in no mood to fund apparent luxuries for all of Greater Victoria. One wonders if they even knew which airport they were being asked to buy. Calling the defeat "smothering," the *Colonist* glumly editorialized on June 30: "The result of the by-law vote indicates that the citizens do not wish Victoria to establish a municipal airport. There is no other conclusion to be reached."

Now the option to buy Lansdowne had also expired, if that made any difference. On August 31, 1931, the City abandoned Lansdowne Field, receiving a release of liability from Mrs. Scott and returning control to the Eve brothers, granting them the pre-paid September rent as a pacifier. Complaints of low and dangerous flying had been received by Dominion authorities, and Lansdowne Field was not to be used for any commercial purposes.

What is remarkable is that both Gordon Head and Lansdowne continued to be viable airfields, at least for the next two years. Lansdowne was very much alive, and Gordon Head's entire short history still lay ahead.

National Air Photo Library, A4517-57

By a happy coincidence, official aerial photographs of Greater Victoria were taken on September 3, 1932, just two days before a major air show at Gordon Head. Because the airfield had been carefully groomed, the sand-and-gravel runway paths are clearly visible.

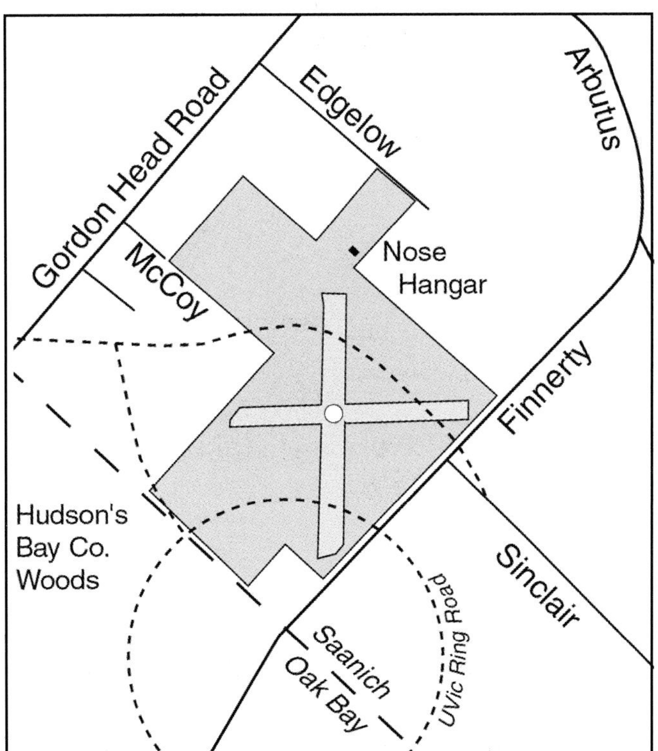


The field lay on cleared farm land in Saanich, between Gordon Head Road and Finnerty Road. Shown here as dotted lines are the modern UVic Ring Road, McGill Road, and McKenzie Avenue. The little nose hangar can barely be seen in the aerial photograph.

Never again after the Trimotor crash did B.C. Airways send passengers aloft, but the Company made a valiant effort to survive the catastrophe of August 1928. The problem was not merely financial; there was also the need to rebuild public confidence in land-based flight over water.

From autumn 1928 to summer 1929, at least the flying school could still operate effectively under Hal Wilson and Ted Cressy. Some of its finest students undertook their training in this period, including whiz kid Maurice McGregor (page 67). However, there was now a burden of debt from the lost Trimotor, and a pall of gloom must have made new initiatives very difficult. What is more, the stock-market crash in the fall of 1929 had a devastating effect on civil aviation.

Both Eaglerocks, G-CAIS and G-CATN, served until the summer of 1929, but they were sold to the Aero Club of British Columbia when that Club lured Hal Wilson to Vancouver on July 24, 1929, to become its instructor. Ted Cressy also departed late in the same summer to accept a position with the new Alaska–Washington Airways. From then on, B.C. Airways really existed only on paper. With Victoria Aero Club for all purposes now defunct, the future of aviation in the city looked bleak.

Still, both Ernie and Cecil Eve were now competent pilots, and flew planes of their own from Lansdowne and later from Gordon Head. For a short time, Victoria could actually boast two reasonably good small airfields.

To close the book on B.C. Airways, we can focus on February and March 1931, when a last-ditch effort was made to revive a moribund but worthy cause. The Company had often been wooed by aviation firms that wished to sell it a successor to its lost flagship. The pictures on this page show two different Tri-motors that came a-courting: a familiar Ford (NC-4532) from Western Pacific Air-freighters, and a Kreutzer (bottom left and top of page). The directors decided to try again, using 6-passenger Kreutzers, linked with Pacific Airways & Steamship Company of Seattle.

A Kreutzer K-5 Air Coach is seen left top, painted in B.C. Airways livery and ready for duty. Maurice McGregor, by now a working professional, was being trained to serve as pilot. However, as the newspapers of February 1931 sadly reveal, the project had a false start or two, faltered, and never got under way. The aircraft and pilot were fine, but other conditions were still not right.

## LANSDOWNE FIELD HOLDS AIR SHOW IN MAY 1931

On Monday, May 25, 1931, barely a month before the Gordon Head referendum, a large holiday air show was presented at Lansdowne Field. Six aircraft flew over from Vancouver, joining two from Victoria. The local planes were Gypsy Moth CF-ADY flown by Maurice McGregor (see photo of Cecil Eve below) and the Eve brothers' Northwest Low-wing CF-AOX, next page. A crowd of several thousand enjoyed precision flying and aerobatics. The eight machines on display formed the largest group of aircraft ever brought together so far in Victoria.

On that Victoria Day holiday, former Victorian A.H. (Hal) Wilson was pilot of Fleet Model 2 CF-ANF (left).

Other Vancouver pilots were Donald Lawson and T. McKinnon in Fleets; R.L. (Ginger) Coote in Eaglerock CF-AMN; A.E. Bennett in Waco 10 CF-AAQ; and Thomas E. Snelgrove in Barling NB-3G CF-AMJ. Captain T.E. Snelgrove was about to become a Victoria resident and a major force behind Gordon Head Airport.

Always an enthusiatic flyer, Cecil Eve (right) strikes a pose beside Gypsy Moth CF-ADY, owned by T.H. Jones of Vancouver but flown for much of 1931 by Victoria's Maurice McGregor. Very soon after this air show, the *Victoria Times* reported that McGregor and his Gypsy Moth would be stationed permanently in Victoria for charter work and sightseeing. "Mr. McGregor has established a fine reputation here and on the mainland for his ability as a pilot. He has been flying for the past four years [actually, two and a half] on all types of machines and for the last two years has been doing instruction work." *Times* 1931.06.06: 15

In the early 1930s, an appreciative crowd at Lansdowne Field watched an unidentified Indo-Canadian daredevil perform a parachute jump from an airplane—quite likely a first for Victoria. The aircraft is an Aeromarine Klemm.

Above, the brand-new CF-AOX awaits delivery in Seattle. Below, it faces the former B.C. Airways hangar, during the air show held on Monday, May 25, 1931. On the day before, Ernie Eve had flown the plane to Vancouver with his daughter Nan, and returned in formation with the six Vancouver aircraft.

B.C. Aviation Museum

When B.C. Airways ceased to be a commercial airline, Ernie and Cecil Eve bought in Seattle a Northwest Low-wing Model A (CF-AOX), powered by a LeBlond 60hp five-cylinder engine. This was one of only two planes of its type built. It was in this machine that Elwood first experienced the thrill of flight. He had earned the ride with Cecil by thoroughly cleaning out the old B.C. Airways hangar.

Cecil continued flying this orange and cream airplane from Lansdowne for many years; he took young Donald Cameron for a ride on Donald's thirteenth birthday in 1939—one of the last powered flights from this field that can be dated.

Cecil's aircraft was the forerunner to the Aeromarine Klemm, whose design drew heavily on the Northwest. One puzzle about his plane was that it would never climb over 500 feet—until it was discovered that it had a 90hp propeller, and the 60hp engine was not delivering enough revolutions per minute. Abe West carved a new prop, and from then on it performed very well.

CF-AOX was finally sold in Chilliwack after Cecil gave up flying.

From CF-AOX, Cecil Eve photographed Lansdowne Field (right) in 1931. We are looking east over Scott Street and Shelbourne Street toward Richmond Road, with Foul Bay Road in the distance and Oak Bay in the upper right corner. Bowker Creek is clearly visible as an oblique dark line, centre right.

Here we see the Waco and Barling aircraft that visited Lansdowne Field from Vancouver for that air show in May 1931. In the top picture, Maurice McGregor stands beside Waco 10 (CF-AAQ), which he flew in 1930 with Sprott-Shaw School of Aviation in Vancouver. The two photos below show the Barling Low-wing CF-AMJ, also a Sprott-Shaw aircraft, which T.E. Snelgrove brought with him when he moved to Victoria on June 4, 1931, planning to operate his own teaching school out of Lansdowne or Gordon Head Field (*Times* 1931.06.05: 11).

Directly below, Victorian Pete Alexander (a future professional) poses with CF-AMJ, in which he took his first solo. Below right, Pete stands beside Gordon Bulger (right), who became Maurice McGregor's air mechanic with Canadian Airways. On the lower half of the page is a story that Maurice loved to tell about Gypsy Moth CF-ADY and James Gray, whom we met in Chapter 4.

"In 1931 [June 10] a former First War pilot, Jimmy Gray , told me he'd like to take a refresher course and fly again but he had some reservations. He was married, had a family and knew that flying was a risky business.

"I told him, 'You're all wrong. Take my word for it. This is 1931. It's an entirely new era, not like your day back in 1918. There have been great technological advances. You remember flying an earlier de Havilland aircraft of the Great War but I have the very latest de Havilland model, a brand new Gypsy Moth with a four cylinder Gipsy engine—totally reliable.'

"Jimmy said, 'If you say so, all right.' He climbed into the front cockpit, I climbed into the back and we took off heading south. After climbing to 300 feet I handed over the controls to him. So there we were climbing happily with the wind blowing around our goggled helmets, heading parallel with Richmond Road towards the hospital when suddenly the engine gave a series of violent coughs and stopped dead. The only sound was the wind whistling in the wires. I called down the speaking tube, 'You'd better let me take her, Jimmy.'

"Victoria fortunately wasn't as built up then as it is today and there were still a number of open lots. I picked one which I could just reach in a glide, sideslipped past the sitting room window of one house, and put the plane down cross-wind close to the back fence of another.

"We came to a stop in dead silence. There was a pause and Jimmy said, 'I thought you said flying had improved. That's enough for me!'

"We climbed out, folded back the wings, and ignominiously pushed the plane back up Richmond Road along the gravel roadbed of the streetcar tracks to Lansdowne Field."

—Maurice McGregor, as told to John Schaffter (*Colonist Islander* 1989.03.26: M2)

## Charles and Anne Lindbergh Take Off from Lansdowne Field

Thursday, October 22, 1931, 4:00 a.m. Charles Lindbergh and his wife Anne, on a world tour, arrived in Victoria aboard the liner *President Jefferson*. Having learned earlier about the death of Senator Dwight F. Morrow, Anne Lindbergh's father, the couple had decided to end their tour and fly home to Englewood, New Jersey, as soon as possible.

The original plan had been to fly east from Seattle when the ship docked in that city. *Victoria Times* city editor Archie Wills (below) had a better idea; three years before, he had made a similar suggestion to round-the-world adventurers Mears and Collyer. Wills informed Lindbergh's friend, pilot Vance Breese, that Victoria had an airport able to provide a more timely connection for a flight across the U.S.A. When the *President Jefferson* landed at Rithet's Wharf, Breese—who had scouted the airfield in advance with Vern Gorst—was waiting at Lansdowne Field with a Lockheed Vega to meet the couple. Lindy himself took the controls at 6:22 a.m., in a thrilling takeoff witnessed by a huge crowd of awestruck Victorians. Many had slept in their cars all night.

The *Times* was able to feature the whole story that Thursday evening, thus scooping the morning *Colonist*. Colonel Lindbergh thought that Victoria's strategic location on the Pacific coast made it an ideal site for a major airport. He even saw excellent potential in Lansdowne Field, though he was not entirely complimentary about its condition: "I have landed a ship and taken off in many fields that have been in worse shape than this. There is plenty of room and the runway, properly leveled off, would be an excellent one."

The *Times* editorial that evening pointed an accusing finger at Ottawa: how could the Department of Civil Aviation obstinately refuse to license a field that the "Lone Eagle" had pronounced satisfactory? Mr. Plunkett, Victoria's M.P., should take up the issue vigorously.

All this time, predictably, Vancouver was fuming. Major Donald MacLaren (Canadian Airways) and Squadron Leader Earl MacLeod (RCAF) had each wired the *President Jefferson*, offering to come across to Victoria by flying boat and escort the famous couple to Vancouver, thus allowing them to depart from a proper airport. Canadian Airways pilot Walter Gilbert, Lindbergh's host on a recent visit to Aklavik (see caption), had also wired Lindy from Vancouver to warn him that Victoria had no licensed airfield.

Photo courtesy Eileen Wilson

Archie H. Wills

Aklavik, NWT, August 1931. En route from New York to Alaska, two months before his Victoria visit, Charles Lindbergh (above left) enjoys a laugh with Canadian Airways pilot Walter E. Gilbert, as air engineer Louis Parmenter looks on. Awarded the Trans-Canada (McKee) Trophy for 1933, Walter Gilbert appears elsewhere in this book as a pilot of Canadian Airways Dragon Rapide *City of Victoria* in Esquimalt Harbour. (*Canadian Airways Bulletin* 5.9 (March–April 1934): 24.)

On October 22, the *Times* Social Page ran a long profile of Anne Morrow Lindbergh. Gracious, fearless, shy, intelligent, and charming, she was an ideal match for a global hero. Her greatest desire was to get back to her little son Charles A. Jr., now 15 months: "When we left home, he was just starting to walk and stand on his feet."

Not five months later, on March 1, 1932, little Charles would be the victim of the most infamous kidnapping-murder crime of the twentieth century.

Anne Morrow Lindbergh

## Gordon Head Becomes Victoria's Main Landing Field ▪ 1932–1933

During the summer and fall of 1931, Tom Snelgrove and his former Sprott-Shaw student, Maurice McGregor, used Lansdowne Field as their base for flight instruction, operating both Snelgrove's Barling CF-AMJ and McGregor's leased DH Moth CF-ADY.

Barely twenty years old, Maurice now established his own company—Vancouver Island Airways, offering flying lessons, charter service, and joy rides. His barnstorming activities outside Victoria were wide-ranging: Brethour's Farm in Sidney, the Menzies Farm on North Pender Island; the Golf Course on Saltspring; the James Seed Farm in Duncan; Grauer's Farm near Boundary Bay; and a vacant Field at Courtenay.

Both pilots were forced to abandon Lansdowne in the fall of 1931, when public pressure caused that field to lose its licence as a commercial aerodrome. Snelgrove also suffered the loss of his Barling, a write-off after a crash at Douglas Lake on the B.C. mainland. McGregor began using the new Gordon Head Airport on November 22, 1931, after Snelgrove had joined forces with Ernie Eve to create British Columbia Aircraft, Ltd., which initially had full responsibility for Gordon Head.

Young McGregor was scarcely ready to be included as a partner, but his services and his DH Moth were critical in the creation of a new flying school. Among the students who would receive flight instruction in 1932 was a group from China, motivated by anxiety about impending war with Japan.

This Gordon Head Airport, now in the heart of the UVic campus, is best remembered for two popular air shows presented by the British Columbia Air Tour on July 4 and September 5, 1932. Its closure in 1933 signalled the end of land-based flight in Victoria for over a decade.

F.M. McGregor Collection

B.C. Airways' hangar stood for years amid the Lansdowne haystacks, a derelict reminder of past glories.

One other event of note occurred at Gordon Head Airport in 1932. This was a special Saturday celebration for young Skyroads members, held ten days before the September air show. As a spinoff from a currently popular American comic strip, the *Victoria Times*, which carried the strip, had been encouraging boys and girls to join the local squadron of a "Skyroads Flying Club." On the basis of written examinations, they could move up through RCAF ranks from Pilot Officer to Wing Commander. We must remember that this was still only five years after Lindbergh's solo flight, and enthusiasm for aviation appeared to be as strong among young Canadians as it was in the United States.

Named as Honorary Colonel of the Victoria Skyroads Club, Maurice McGregor (now a seasoned veteran at 21), suggested that he could give the children some practical experience. So it was that he hosted a gathering of young Club members at Gordon Head Airport on Saturday, August 27, 1932. The Skyroaders were transported to the field in two large trucks belonging to Dowell's-Pacific Transfer Company Limited. Once there, they were all offered rides on a cent-a-pound basis—one dollar minimum (an enterprising formula).

The aircraft was the small two-place Gypsy Moth (CF-ADY) that Maurice had flown in Victoria for most of 1931; it was still stationed with Tom Snelgrove at Gordon Head. Though the whole procedure might give fits to a modern parent, the *Times* heartily approved:

> The Skyroaders got their greatest thrill out of the expert manner in which 'Colonel' McGregor handled the little moth plane. Hardly touching the field for more than ten seconds at a time, he changed passengers in record time, taking off and landing thirty-seven times in less than two hours . . . .
>
> By actual count, thirty-six boys and one girl took flights, the girl being Rubymay Brown [later Parrott] . . . who is a flight lieutenant in the club. Rubymay's two brothers, who are also Skyroaders, went up as well, and their mother came along to watch them.
>
> —*Times* 1932.08.30: 7

Dr. Harry M. Evans, former Registrar at Simon Fraser University, remembers as a young Victoria Skyroader receiving his first plane ride on a similar occasion at Esquimalt Harbour, in a pusher flying boat.

J. ERNEST EVE
Phone E 2117

OFFICE: 900 FORT STREET
Phone E 4168

T. E. SNELGROVE
Phone E 5657

BRITISH COLUMBIA AIRCRAFT, LTD.

*Controlling and Operating*
SAANICH AIRPORT

VICTORIA, B. C.

UVic Archives 168.1101 (Daphne Cariou)

Perhaps the man most saddened by the defeat of the June 1931 airport by-law was Captain Thomas Eyre Snelgrove, ex-Royal Flying Corps and former chief pilot of Vancouver's Sprott-Shaw School of Aviation (in trench coat and tam above), who had moved to Victoria that very month with the plan of starting another flying school. Not one to give up on a good cause, he proceeded in late September to lease and license the land that the City had assembled, and Gordon Head Airport came into being. With support from Ernest Eve and Maurice McGregor, his school continued at Gordon Head.

The only aircraft kept continuously at Gordon Head was a Fairchild KR-21 Taperwing (CF-AMH), owned by Bob Proctor of Victoria (far right, top photo). It was housed in a nose hangar (above and page 80) off Edgelow Road.

Very likely Tom Snelgrove had over-extended himself in leasing Gordon Head Airport, as the City of Victoria received a letter in October 1931 in which he requested financial help with improvements. On October 30, the Industrial Committee moved that "Mr. Morton be authorized to arrange for the necessary marking of this Field to assist visiting aircraft, at a cost not exceeding $50.00." This was the thin edge of the wedge. On April 21, 1932, the City of Victoria was drawn into airfield operation yet again, when its committee was told by Mr. Carter Guest that the Gordon Head licence would be cancelled unless certain improvements were made. "Alderman Straith stated that Major [*sic*] Snelgrove had leased certain land for the Winter but had defaulted in his payments of rent."

A motion was passed to lease all necessary property, and to place $1,000 in estimates for airport purposes (rental and other needs). Saanich Municipality would be asked to put an amount equal to taxes—about $500—into runway improvement.

Under a new mayor (David Leeming) and a new Chairman (Alderman Todd), the City's renamed Business and Trade Development Committee was in for another year and a half of airport management.

UVic Archives

UVic Archives

Because Eve Bros. held the Victoria franchise for Alexander aircraft, Ernie Eve placed his professional stamp on this Company brochure for the new (1931) Alexander Flyabout, "the first practical flivver plane." In 1928, B.C. Airways had acquired two Alexander Eaglerocks (G-CAIS and G-CATN), but it seems that no Victorian other than Ernie bought another Alexander aircraft. The Flyabout was registered as CF-ASX on June 14, 1932, when Lansdowne had been almost entirely abandoned in favour of Gordon Head Airport—where these three photos were taken. Shown above (on right) with Ernie Eve, Maurice McGregor flew this machine all over the Province during the B.C. Air Tour in the summer of 1932. Sharing ownership of the airplane in 1936 with B. Low of Victoria, Ernie kept Flyabout CF-ASX until 1940.

Here Maurice McGregor is photographed before the B.C. Airways hangar at Lansdowne Airfield with the Alexander Flyabout aircraft (CF-ASX) owned by Ernie Eve. It was this airplane that he regularly used in his celebrated "Crazy Flying" routine, as performed at Gordon Head Airport on September 5, 1932.

Maurice McGregor Collection

Maurice McGregor had already qualified for his commercial licence at age 18, but it did not become valid until his 19th birthday on May 11, 1930. For the next three years, until he joined Canadian Airways, he spent much of his time as a free-lance entrepreneur, establishing and operating Vancouver Island Airways out of Lansdowne and Gordon Head, with Gypsy Moth CF-ADY.

In the summer of 1932 he became quite famous as a star performer with the B.C. Air Tour, which appeared twice at Gordon Head Airport: July 4 and September 5 (Labour Day). Maurice was one the Three Musketeers, an excellent aerial precision team. But he was best known for a solo act called "Crazy Flying," which he performed always in Flyabout CF-ASX. John Schaffter, who wrote several elegant profiles of Maurice McGregor, has given us the best description of this piece of aerial melodrama. We reproduce it with his kind permission.

Shows would end with the announcement of flights for members of the public. To get things started a favourite trick was one first performed at the Cleveland Air Show in Ohio by Squadron Leader Achterley of the RAF, then holder of the world's airspeed record, and introduced to B.C. by Maurice McGregor.

A couple of mechanics would wheel a plane out onto the field and the master of ceremonies would announce through a bull horn the start of $1.50 joy rides.

"The first person to ride is going to be Granny Jenkins. She's 87 years old and not in the least afraid. So what's the matter with you young fellows? You're not nervous are you? Come along now, line up for your flight."

The crowd watched as a pilot with a leather helmet and goggles escorted Granny out to the aircraft. She had snow white hair, a little black bonnet, and a black dress down to the ground. She was obviously frail. The pilot gallantly helped her up into the open cockpit, then climbed in beside her and started the engine.

Just as the plane was about to take off, a mechanic with a clipboard ran out waving his arms and gesturing. He obviously had an urgent message.

The pilot climbed out, leaving Granny alone for a moment in the cockpit with the engine ticking over. She must have bumped the throttle for the plane suddenly shot forward. Veering wildly from side to side, it headed for the thickest part of the crowd. The spectators were petrified, especially those in line with the spinning propeller.

At the last moment, the plane spun around in a sharp half circle, one wing tip almost striking the ground, then veered away, still wobbling erratically, and gathered speed. There was a gasp from the crowd as the aircraft suddenly nosed up into the air, climbed a few hundred feet, hung for a moment on its propeller, then stalled and fell spinning toward the ground. Fortunately, Granny Jenkins (Maurice McGregor in drag) was able to pull out of the stall just before the wheels hit the ground and bounced the aircraft heavily back up again into the air. Low flying followed, with one wing tip dragging inches off the ground to throw up a swirling cloud of dust.

—John Schaffter, "Maurice knows joy of flying," *Times-Colonist Islander* 1991.05.12

## Victoria's William H. Cross and Jarl "Grubby" Grubbstrom

One of Maurice McGregor's earliest students of the 1930s was Victorian William H. (Bill) Cross. He took to the air like a bird, the challenge intriguing him. In 1931 he went to New York to receive extensive instruction on heavier aircraft. Meeting up with Jarl Grubbstrom, a young Swede who had logged some 50,000 miles of flying experience in Eastern Canada, Cross purchased from Canadian Airways a Fairchild FC-2, registered as G-CAIH. Then he and "Grubby" Grubbstrom made an historic winter flight across Canada from Cartierville, Quebec, to Victoria's Lansdowne Field.

Leaving Quebec in January 1932, after a delay caused by snow, they landed the machine at Lansdowne Field on February 23, having recorded only 28 hours of flying time. Several problems beset the flyers en route. Once they plunged through the ice in a slough at Creston, B.C., breaking the propeller. Stranded there, they had to wait three weeks for a new prop to be shipped from the Fairchild factory in Quebec.

The adventuresome pair and their trusty Fairchild were to be seen often in B.C. skies the rest of that year. On the 1932 B.C. Air Tour circuit, the plane was used not only to tote belongings for the crews of the smaller aircraft, but also to provide a platform for the parachute jumping exhibition. G-CAIH was eventually sold to Pacific Airways of Vancouver, and later rebuilt as a Fairchild 51. While serving on the Zeballos run as Ginger Coote Airways CF-AUX, she went down fifteen miles west of Port Alberni on May 27, 1938, and was not found until March 9, 1939 (see Schofield 2004: 16–23).

Fairchild G-CAIH at Lansdowne Field in 1932.

Bill Cross was the hero (or anti-hero) of a popular story that came out of the B.C. Air Tour's visit to Victoria on September 5, 1932 (photo above). It concerned a parachute jump he made over Gordon Head Airport that day from his own Fairchild, G-CAIH.

Until then only an observer of other daredevils on the Tour, Bill decided to take his first jump—but then had second thoughts about the whole idea. To calm his nerves, his friends took him to the nearby home of M. LaRivière, famous for his home-made cherry wine.

The wine seemed to work, but Bill again lost his nerve when the crucial moment came. His friends helped him decide—with a swift push out the aircraft's open door.

In addition to the aircraft that were officially part of the 1932 travelling B.C. Air Tour, several private machines appeared for the Gordon Head Air Show. Among the more unusual was King Baird's Autogiro, NC-11608.

The B.C. Air Tour first performed at Gordon Head Airport on July 4, 1932. The aircraft lined up above include three Fleet Model 2 biplanes, powered by Kinner K-5 engines, flown by the so-called Three Musketeers: CF-ANL, CF-ANF, and CF-ANN. One of them is taking off from Gordon Head in the photograph at the bottom of the page.

UVic Archives

Both air shows were very well attended. Because of major traffic problems on July 4, the Labour Day meet on September 5 was better planned and organized.

Three Fleet Model 2 aircraft formed the nucleus of that B.C. Tour. CF-ANL and CF-ANN, acquired in 1930 by Vancouver's Aero Club of B.C., were the first and third planes manufactured by Fleet Aircraft of Canada Ltd., Fort Erie, Ontario. In the photo at left, Hal Wilson (in white) stands beside ANL, named *Austin C. Taylor* after Home Oil President and Aero Club sponsor.

CF-ANF, seen (left) near the Gordon Head nose hangar, was first registered to Fleet Aircraft of Canada, having been made in the U.S. by Major Reuben H. Fleet's Consolidated Aircraft Company. In Vancouver, its owners included Commercial Airways (1930), A.H. Wilson and W.G. Mackenzie (1932), and L. Foggin (1937), who was killed when he crashed the plane near Boundary Bay on May 13, 1941.

## Speed Run and Precision Flying at Gordon Head ▪ Summer 1932

At both of the 1932 Gordon Head Air Shows, there was a speed demonstration put on by the latest model of Lockheed Vega, a plane renowned for its swiftness—almost 200 mph. The aircraft, seen at left and in flight directly below, was Shell Oil No. 5, flown by Bill Fletcher as a promotion for Shell Aviation Company of San Francisco. Because it is wearing "pants" below, one occasion must be July 4 and the other September 5.

City of Victoria Archives PR19-6847

City of Victoria Archives PR19-6844

Now beginning to break formation in their run above Gordon Head Field, these three Fleets are being flown by the celebrated trio known as The Three Musketeers: Hal Wilson, Don Lawson, and Maurice McGregor. Though far, far slower than modern precision counterparts, their aerial manoeuvres were no less complicated.

For the nostalgic reader, the parked automobiles seen here may hold as much interest as the airplanes.

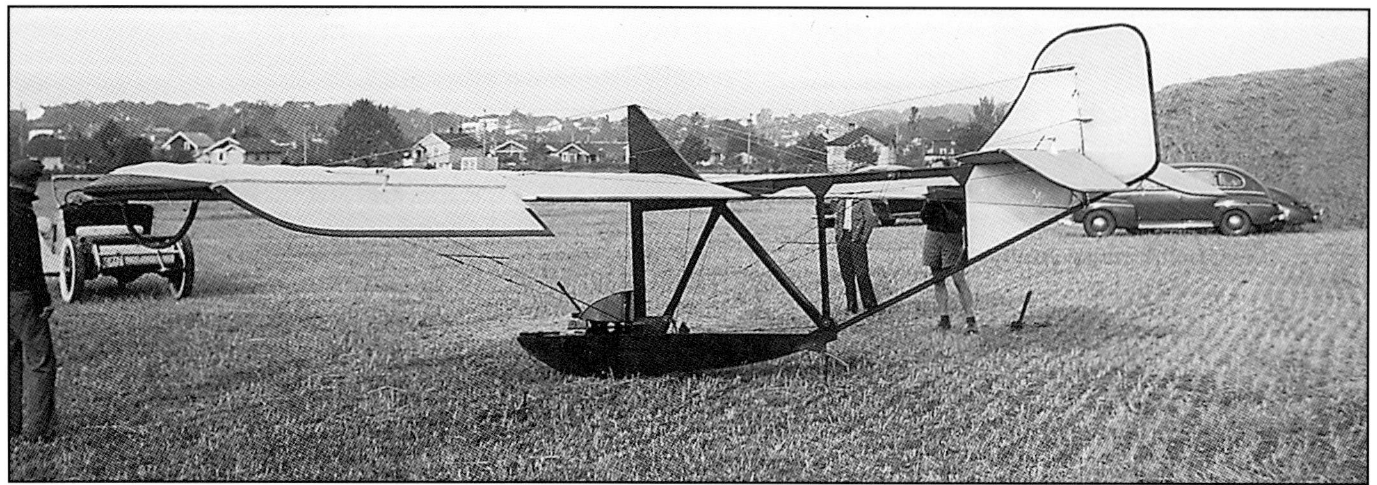

Gliding is a form of aviation that provides an abundance of thrills in an atmosphere of quiet serenity. A Victoria Glider Club was formed at Lansdowne Field about 1930.

Experienced pilots like Nick Carter and Jimmy Gray are reported to have reached heights of 400 to 500 feet. Major-General J.H. MacBrien, on a visit to Victoria, was highly complimentary: "This is the best gliding I have seen in Canada" (McKibben 1979: 4). In that period, the machine used was a primary (beginner's) glider, like those seen at Lansdowne Field and Gordon Head Airport in the lower two photographs.

In 1942, the Victoria Gliding and Soaring Club was formed, with Alderman B.J. Gadsden and J.B. Taylor as president and vice-president. In the upper two pictures, we see a Northrop glider, under construction in 1943, and, at the top of the page, in use at Lansdowne Field. A Dagling (see next page) was built in 1944, but it crashed at Lansdowne in 1945. A Kirby Cadet was also built and flown in 1945. After this machine, too, was seriously damaged in a crash, the Club soon disbanded.

In 1956 a third Gliding Club was formed, acquiring in Seattle a Swetzer two-place machine. This was extensively rebuilt at Fairey Aviation (see Chapter 25). Eventually, however, the Swetzer was bought outright by a few members, and then was flown mainly from Cassidy Airport, south of Nanaimo.

Gliding was performed in the early 1930s at Gordon Head Airport (see photograph at right). The Gordon Head Air Show of September 5, 1932 opened with an exhibition by Ewen Boyd and Eric Mahon in a primary-type glider, launched by being towed behind an automobile (*Times* 1932.09.06: 11).

UVic Archives 168.1102 (Daphne Cariou)

W.W. Taylor Collection

W.W. Taylor Collection

From the photographs above, we gain some understanding of how the Victoria Gliding and Soaring Club normally launched its gliders at Lansdowne Field in the 1940s. There was a very ingenious albeit primitive technology, which must have been thrilling to experience and exciting to observe.

In two "official" vehicles, Club members might arrive at the northeast corner of Lansdowne Field—recognizable above by the familiar Normal School building beyond Richmond Road. A 1928 Packard roadster pulled a trailer that contained a glider such as the Dagling (seen here beside a large haystack); assembly was required prior to takeoff. When the glider was ready for flight, at the highest point of the field, the Packard would drive to a distant point south of Bowker Creek, where it assumed the role of a motorized but stationary winch. The tow rope (or metal cable) was extended to the glider from a winch axle that joined a pair of friction-driven automobile tires positioned above the Packard's own two rear tires.

When the Packard was jacked up so that its rear wheels were off the ground, and the engine run in high gear, the glider could be drawn down the hill faster and more effectively by winch than if it were merely towed behind another vehicle in the old-fashioned, uncreative way.

The pilot of the glider controlled a mechanism to release the tow cable at the appropriate moment, when he was already soaring high above the field. Of course, the speed of the winch and timing of the release would depend upon the skill and experience of the pilot.

The second "official" vehicle was the stripped-down Model T Ford that is seen in two photographs above. Its role was not glamourous, but was very practical: it served as a motorized dolly on which to carry the glider, after its down-field landing, back to the starting point for another winch-assisted flight.

Desmond J. Cavin describes his gliding adventures in "Panhandle Pilot," a 1989 reminiscence (Cavin 1989). He notes that the glider might be as high as 600 feet in the air before it was time to pull the release handle. He also points out, logically enough, that the direction of towing would depend on the wind conditions of the day.

Home-built by Victoria garage mechanic Stan Anfield (right), Pietenpol CF-AOG was successfully test flown at Gordon Head Airport by Maurice McGregor on February 17, 1933, and licensed on May 16, 1933. Powered by a 50hp Model A Ford engine, it had a top speed of 85mph and a cruising speed of 70. In this 1933 photograph, Mount Douglas is in the background.

In the picture below, also taken in 1933, Ernie Eve stands alongside Pietenpol CF-AOG. Behind the plane is the Gordon Head nose hangar.

In 1967, John Howroyd decided to rebuild the derelict Pietenpol, encouraged by his aviator father Joe. Eight years of hard work later, on July 8, 1975, he registered the vintage CF-AOG in his own name. On April 12, 1986, at age 75, Maurice McGregor was taken for a ride by John Howroyd in the aircraft that the young Maurice had test flown 53 years before. Still flying in the 21st century, the rebuilt Pietenpol is seen below at Quamichan Lake.

John Howroyd Collection

## PART FOUR

# BY SEAPLANE TO VANCOUVER AND SEATTLE

## 1929–1945

# 10. Alaska–Washington Airways ▪ 1929–1931

In the twelve months after the Ford Trimotor crash on August 25, 1928, there was no scheduled commercial air traffic between Victoria and the mainland, though there was much discussion about an air harbour. Victoria City Council and the Chamber of Commerce both agreed on the need for such a facility.

It was apparently assumed that some outside interest would restore the severed link, now that B.C. Airways had fallen on hard times. Indeed, when Winnipeg's Western Canada Airways merged with Vancouver's Dominion Airways in 1929, that company announced a plan to provide Victoria with twice-daily service, but nothing more than special flights ever materialized.

To the left we see a Western Canada Airways flying boat (Boeing B-1E G-CAUF) at the JBAA wharf in the Inner Harbour. The photo can be fairly precisely dated by the state of progress on the Empress Hotel's new North Wing: the occasion is very likely a well-publicized flying visit to Victoria by Manitoba Premier John Bracken on July 24, 1929.

The second picture is very close in time; it shows an Alaska–Washington Airways seaplane (Lockheed Vega NC-657E, named *Ketchikan*) at the same wharf. The date is almost certainly August 13, 1929: on that day, in another well-publicized visit, senior officers of Seattle's Alaska–Washington Airways were flown to Victoria by former local boy Alex Holden, for a meeting in the Empress Hotel. The one item of business was the creation of a Canadian subsidiary, Alaska–Washington Airways of British Columbia, Ltd., aimed at providing regular seaplane service between Victoria and the mainland. Like B.C. Airways, this was to be a Victoria-based operation.

The bottom picture shows the same aircraft in the same position, and presumably on the same day. The original *Princess Marguerite* is at the CPR dock across the harbour.

Six days after that Empress Hotel meeting, Alaska–Washington Airways inaugurated its Seattle–Victoria service with a VIP flight on Monday, August 19, 1929. The seaplane was a Fairchild 71 (U.S. registration NC-115H, already bearing the name *Victoria*), flown by the veteran American pilot, Captain Gordon K. MacKenzie.

This and all later revenue flights used the B&K wharf at Shoal Point (photos at right). Regular service began Tuesday, August 20 with twice-daily Victoria–Seattle round trips; these left Victoria at 9:30 a.m. and 4:30 p.m., returning from Seattle at 11:00 a.m. and 5:30 p.m. The fare was initially $13 one way and $25 round trip.

The B.C. subsidiary under H.B. (Barney) Olson was not official until September 19, and the start of Victoria–Vancouver service was delayed until October 16.

Both Victoria newspapers gave detailed coverage to the inaugural Seattle flight on August 19, 1929. The poor-quality image below was copied via microfilm from a front-page photograph in the *Victoria Times* on August 20. That morning, the *Colonist* had run a vivid first-hand description of the return flight to Seattle, along with a delicious story about George (Rebel) Mowat, then with Grey Line Tours. When Captain MacKenzie carelessly dropped his set of keys into the harbour, the hefty Mowat, a legendary Victoria character, at once stripped off most of his clothes and dived off the seaplane at the B&K float, descending into the icy depths twice—alas, with no success.

Capt. Gordon MacKenzie (left) would fly this aircraft, later CF-AJP, for two British Columbia airlines.

Aboard the inaugural flight from Seattle (L–R): H.B. (Barney) Olson, head of AWA B.C. Ltd.; Mrs. J.L. Carman Jr., wife of AWA's president; Mrs. H.F. Alexander, her wealthy Seattle mother; Captain Gordon K. MacKenzie, pilot; Frances Frink of Seattle; and Alex Holden, now AWA's vice-president.

Just four months before the inauguration of Victoria–Seattle service, Joseph L. Carman had established Alaska–Washington Airways with a pioneer 940-mile non-stop flight from Seattle to Juneau. The date was April 15, 1929; the aircraft was a Lockheed Vega (NC-432E) named *Juneau*—Carman liked to give city names to all his AWA fleet. Seattle-born Anscel Eckmann was pilot of that historic flight to Juneau; Robert E. Ellis was navigator; and Jack Halloran was mechanic. Thus was born a fabled tradition of commercial flying in the Alaska Panhandle.

Joseph L. Carman was a Seattle furniture manufacturer turned aviation entrepreneur. He had previously opened a small flying school, and was doing charter work in the Northwest with a Stinson Detroiter and a Lockheed Vega. His father-in-law, H.F. Alexander, was president of the esteemed Admiral Steamship line of Seattle; he gave Carman financial backing and encouragement to expand into Canada and Alaska. The pair were indeed visionaries for their day: the ultimate goal was an air route to the Orient (Satterfield and Jarman 1969: 23).

Carman's Canadian subsidiary was to be known as Alaska–Washington Airways of British Columbia, Ltd. President was H.B. (Barney) Olson, a Victoria entrepreneur who is profiled at the bottom of this page. Directors would include the wealthy Victoria industrialist Robert Pim Butchart (of Butchart Gardens fame) as well as Joseph L. Carman himself. This was clearly a very solid venture; the only bad omen was the fact that the Canadian service was launched in October 1929, month of the now infamous stock-market crash. The subsequent Great Depression was a major cause of Alaska–Washington's disappointingly brief life-span, both in Canada and in the U.S.A.

The two photographs on the left show AWA Fairchild NC-115H in moments of repose at Shoal Point.

Courtesy of Grant Olson

Born in Milltown, Wisconsin, Barney Olson was a 16-year-old farm boy with four dollars in his pocket when he came to Victoria in 1912. Pursuing his passion for cars (see left), he soon was chauffeur to the manager of VMD. In 1918 he became part-owner of a taxi and sightseeing company that he built into a Vancouver Island transportation empire. His brief romance with aviation was a rare unprofitable venture. His ice rink at Willows Fairgrounds operated from 1941 to 1944. Today he is best remembered for his ownership and creative development of the Strathcona Hotel, which is now a third-generation Olson family enterprise.

Directors of Alaska–Washington Airways were back in Victoria on Thursday, September 19, when final steps were taken to establish the British Columbia subsidiary. The morning *Colonist* reported that Victoria–Vancouver service was expected to begin in four days, on Monday, September 23, using a seven-place 425hp Fairchild (the now-familiar NC-115H). The Seattle route would be served by a Lockheed Vega seaplane capable of 165 mph under full load. That Vega was identified as the record-setter in April 1929 between Seattle and Juneau (*Colonist* 1929.09.19: 17).

The next day's papers reveal that H.B. Olson was elected president, with Captain Alex B. Holden vice-president in charge of operations. The *Colonist* adds other details: Captain Gordon MacKenzie would now fly the Fairchild on the Vancouver run, while Mr. Floyd Keadle (with 2,700 hours flying time since 1912) would be pilot of the Seattle Vega (*Colonist* 1929.09.20: 6). That very day Holden was flying to Vancouver, to confirm all these plans with Squadron Leader MacLeod.

September 20 was not going to be a great day for Alex Holden. No doubt tactfully, S/L MacLeod must have reminded him that Canada had firm regulations about the licensing of aircraft and pilots. Since the Fairchild was now going to be owned by the Canadian subsidiary, it would be a simple matter to re-register it in Canada and paint over the U.S. numbers; but those steps must be taken. It was perfectly legal, of course, for U.S. pilots to fly into Victoria from Seattle on an international route. However, any pilot flying passengers between Canadian cities must pass a rigorous examination so as to qualify for a Canadian commercial licence. For all his professional experience, Captain Gordon MacKenzie was not licensed to fly between Victoria and Vancouver. (MacLeod probably pointed out that even Vice-President Holden, Victoria born and reared, and a Royal Flying Corps veteran, was not yet licensed for that route.)

Accordingly, we learn that both MacKenzie and Holden now had to hit the books in order to earn their Canadian licences: "Capt. G.K. MacKenzie, A.F.C, and Capt. Alex Holden of Seattle are both in Vancouver today undergoing examination for pilot's licences in Canada" (*Times* 1929.09.24: 1). Procedures were set in motion to register the Fairchild in Canada, and the Vancouver operation was temporarily put on hold. So as to minimize delay, Barney Olson persuaded B.C. Airways veteran A.H. (Hal) Wilson to serve as pro-tem reserve pilot on the Vancouver run. Agreeing cheerfully—he had a job with the Vancouver Aero Club, but still yearned to be an airline pilot—he underwent a quick training course on the Fairchild, now registered in Canada as CF-AJP.

Hal Wilson's former Victoria associate Ted Cressy was hired as air engineer on the tenth of October. Over sixty years later, Cressy recalled an eventful first week of employment (White 1995: 8):

> "We flew tests of AJP from Victoria to Vancouver and got our DoT [Civil Aviation] licence for that plane. We had a trip to Comox on the eleventh and on the 12th we returned to Victoria. On the 14th [Monday] we were in English Bay in Vancouver and were hired by Pathe News to fly back to Victoria to photograph the *Empress* shipwreck. The *Empress* liner had run aground and she was right tight to the shore."

The two photographs above depict that grounding of the *Empress of Canada* near Albert Head on October 13–15, 1929—an accident in which the CPR liner sustained serious damage. By a stroke of luck, we see the Fairchild present at the scene, with its new Canadian registration letters, CF-AJP, hastily painted on the fuselage and wing. At the starboard end of the wing, one can detect the old U.S. registration faintly showing through: NC 115[H].

Hal Wilson and Ted Cressy would fly the inaugural Victoria–Vancouver flight on October 16, 1929.

Regular revenue flights between Victoria and Vancouver actually began on Thursday, October 17, 1929. The inaugural round-trip on Wednesday, October 16 was the usual ceremonial event, with civic and aviation dignitaries gathered to celebrate the occasion. Hal Wilson brought the Fairchild smoothly into the English Bay terminal at 1:35 p.m., after a 39-minute flight from Victoria, an event documented in the upper photograph at right. Names of the reception party (right) were listed that evening in the *Vancouver Sun* (1929.10.16: 24). A picture of the Vancouver-bound passengers boarding the plane on her maiden voyage from Shoal Point appeared in the *Times* on Friday (1929.10.18: 14).

Hal Wilson soon got bumped from his new duties: the *Times* lists Gordon MacKenzie as the Fairchild pilot for Sunday, October 20, and he appears in a photograph taken October 25, flying Lt.-Col. Cy Peck, V.C., to Vancouver (*Times* 1929.10.21 & 26).

For the next year and a half, both the Vancouver and Seattle services appeared to run smoothly and efficiently, despite the lack of hangar or maintenance facilities in Victoria. Major repairs could be arranged in Vancouver at the new Boeing Canada aircraft factory (near Stanley Park), and in 1930 Alaska-Washington of B.C., Ltd., obtained hangar and office facilities in the Air-Land float-plane hangar on Sea Island. But because the local air harbour on Erie Street was never built, this Victoria-based company could not keep its own aircraft in town overnight, a circumstance that had a major impact on scheduling.

October 16, 1929: CF-AJP's inaugural scheduled flight is greeted at Dominion Airways Wharf, near the foot of Broughton Street in English Bay. The Vancouver welcoming party included Aldermen Harry DeGraves and E.W. Dean; Vancouver Airport Manager William Templeton; Police Chief W.J. Bingham; and several representatives of Home Oil Company, whose products AWA of B.C. had agreed to use exclusively. Home Oil vice-president Austin C. Taylor stands right centre, arms crossed and holding hat. Pilot A.H. (Hal) Wilson is in the middle, wearing a grey suit and fedora—looking less military than the uniformed Captain Gordon MacKenzie. Engineer Ted Cressy rests his hand on the propeller in the background.

September 13, 1930: 10:45 a.m. As Alaska–Washington of B.C. neared the first anniversary of Victoria–Vancouver service, its Fairchild, flown on this particular occasion by Pat Renahan, made a test airmail run between Vancouver and Victoria, later the same day making a similar flight from Vancouver to Nanaimo. (*Times* 1930.09.13: 1)

Here CF-AJP is docked at the B&K wharf, as Barney Olson (far left) watches postal officials and other dignitaries strike a pose. The aircraft is displaying its distinctive Canadian livery.

*Victoria Daily Times* 1930.01.13: 7

Both the Vancouver and Seattle service used the same primitive terminal at Shoal Point. The volume of traffic made it necessary to build a double floating wharf, along with a crude shelter and an official Home Gas facility. If Percy Barnes arrived with his Boeing flying boat, there could be three aircraft here at once, as reported in the *Times* on October 21, 1929. Here we see Fairchild CF-AJP in its new livery. On Christmas Day, 1929, aided by a brisk tail wind, Captain Gordon MacKenzie flew it from Victoria to Vancouver in a record-setting 22 minutes, earning Home Oil stardom in the advertisement at left—though the graphic artist mangled his name, got the flight direction wrong, and depicted the aircraft with its obsolete U.S. registration.

Alaska–Washington's chief pilot, Anscel Eckmann (above and p. 40) already knew Victoria well, and often flew the cream-and-blue Vega *Ketchikan* on the Seattle–Victoria run. Here are three views of the 5-place *Ketchikan* (NC-657E). Other regular pilots on this route were Alex Holden, Robert Ellis, and Robin (Pat) Renahan. This was likely the aircraft in which Holden set a Seattle–Victoria speed record of 29 minutes (with a full load) on January 2, 1930.

With the help of the map below, we can interpret this photograph and the two across the page, which show operations of Alaska–Washington Airways at Shoal Point. In the picture above, dating from either 1930 or 1931, we are looking over the still-primitive B&K wharf in the direction of Huron Street, where the high ground blocks our view of the streetcar terminus and Rithet's Outer Wharf. There are two aircraft at the double floating wharf: the Vancouver Fairchild (CF-AJP), left, and a Seattle-bound Lockheed Vega (foreground), named *Petersburg* (NC-336H), a red-orange aircraft flown variously by Anscel Eckmann and Robert E. Ellis. Again we see the Home Gas pump. Behind Fairchild CF-AJP is an area previously occupied by Victoria Chemical Company, but now abandoned.

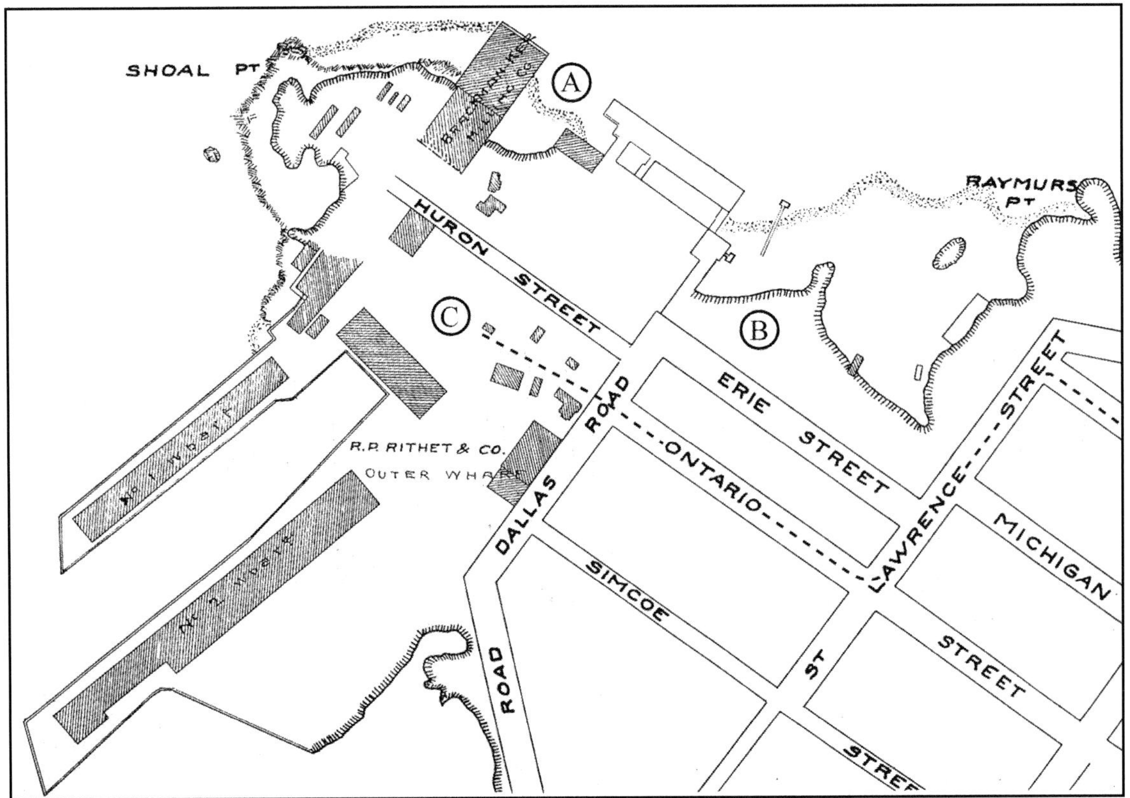


Adapted from Thomas Charles Sorby Map of Victoria Harbour, July 1922 (City of Victoria Archives, D0016)

In the 1920s, Huron Street east of Dallas Road provided access to Rithet's Outer Wharf and to the Brackman-Ker National Mills, where Eddie Hubbard had built the B&K seaplane wharf (A). From 1928 to 1930, the City of Victoria planned to create a proper Air Harbour (B) off Erie Street, where Fisherman's Wharf was later developed. This site was close to the terminus (C) of the Outer Wharf streetcar, the route of which is shown here by a broken line.

City of Victoria Archives PR 92-6868

Having just arrived from Vancouver, Fairchild CF-AJP is taxiing towards the Shoal Point seaplane dock. This 1930 photograph by Victoria teacher Arthur L. Bagshaw provides an exceptionally clear view of the Songhees Industrial Reserve in Victoria West: (L–R) Canadian Western Cooperage, Cooperage Wood Co., Union Oil, Shell Oil, and Sidney Roofing. In the right foreground we see the "very dangerous" old Chemical Company wharf, which the City of Victoria was pressing Canadian Industries Ltd. to remove (Industrial Committee 1930.08.01).

City of Victoria Archives PR 92-6867

Facing northeast, Bagshaw photographed two Alaska–Washington aircraft against a colourful backdrop—derelict remains of the Chemical Company, log booms at Raymur's Point (now Fisherman's Wharf), and distant glimpses of downtown Victoria, including St. Andrew's Presbyterian Church, the Empress Hotel, and the dome of the Parliament Buildings. The Lockheed Vega in the foreground is the new 7-place *Juneau* [II] (NC-103W), soon to be renamed *Skagway*. Piloted by Robin (Pat) Renahan, it disappeared between Prince Rupert and Juneau, with the loss of three lives, on October 28, 1930, during the famous search for missing Canadian pilot Paddy Burke.

The B.C. affiliate of Alaska–Washington Airways appears to have operated for about eighteen months. The official schedule for its second season is reproduced below, but a news story in February 1931 reveals that Victoria–Seattle service was then reduced to one flight a day (*Colonist* 1931.02.13: 15). It claimed that AWA would soon have an 11-place, 900hp Fokker trimotor on each of the Seattle and Vancouver routes, but those ambitions did not match the Company's fortunes at the time. Indeed, the retrospective tone of that article makes it read almost like an obituary. By May 1931 both Victoria services were moribund, if not dead.

There had been such high hopes back in 1929. A great expansion had been announced that year in late December: by the following April, two huge Savoia-Marchetta flying boats, each capable of carrying sixteen passengers, would be operating on the Vancouver–Victoria–Seattle route, and two more would be in service between Seattle and Alaska (*Times* 1929.12.31: 1). At the time, this seemed a reasonable expectation.

Despite good planning and management, AWA began to suffer a series of mishaps and setbacks in early 1930. Gordon MacKenzie severely damaged the Fairchild in a foggy take-off at English Bay, when his right float struck a large buoy. Under Ted Cressy's supervision, CF-AJP was given a major refit, but because there was no backup Canadian aircraft, Vancouver–Victoria service was suspended, and reserve pilot Hal Wilson was once again, as he put it, "crashed out of a job" (White 1995: 7).

Meanwhile, Vega *Juneau* (see top right) was sent to the factory and was judged unfit for overhaul. On August 4, 1930, an almost new 7-place Vega named *Taku* (NC-102W), not normally on the Victoria route and flown by a non-regular pilot, W.A. (Billy) Williams, was forced to land in Puget Sound near Kingston. Though Williams and his five passengers were unhurt, the aircraft caught fire and was totally destroyed (*Colonist* 1930.08.05: 1). This loss was followed in October by Pat Renahan's tragic disappearance in the Vega *Skagway* (NC-103W; Schofield 2004: 43–48). Not only was Joseph Carman losing planes; he was recording operating losses both in Alaska and on his southern routes (Washington and B.C.).

In March 1932, about a year after AWA seaplanes stopped flying out of Victoria, a Seattle receiver's sale sold the American company's assets, and J.L. Carman went back to building furniture. By then, Wells Air Transport of Vancouver had bought Fairchild AJP, after it had completed some 3,000 flights across Georgia Strait.

The first Lockhead Vega named *Juneau* (NC-432E), in which Anscel Eckmann had made his pioneer flight to Alaska on April 15, 1929, was withdrawn from service after only ten months' use. Here it is seen leaving Shoal Point in the shadow of a Canadian National barge that is moored at B&K's National Mills.

For years after his venture into aviation, Barney Olson kept a framed picture of an aircraft on the wall of his Strathcona Hotel. "That piece cost me $25,000," he would tell new friends. "How could that picture cost $25,000?" was the expected reply. "Well," he would laugh, "that's what the stupid plane cost me by the time I was through with it."

AWA

ALASKA-WASHINGTON-AIRWAYS

**Schedules and Tariffs Effective September 27, 1930**

BRITISH COLUMBIA AND PUGET SOUND DIVISIONS

**Seattle—Victoria**

| | | | |
|---|---|---|---|
| Lv. Seattle | 10:00 a.m. | Ar. Victoria | 10:45 a.m. |
| Lv. Seattle | 3:00 p.m. | Ar. Victoria | 3:45 p.m. |

**Victoria—Seattle**

| | | | |
|---|---|---|---|
| Lv. Victoria | 11:00 a.m. | Ar. Seattle | 11:45 a.m. |
| Lv. Victoria | 4:00 p.m. | Ar. Seattle | 4:45 p.m. |

**Vancouver—Victoria**

| | | | |
|---|---|---|---|
| Lv. Vancouver | 10:00 a.m. | Ar. Victoria | 10:45 a.m. |
| Lv. Vancouver | 3:00 p.m. | Ar. Victoria | 3:45 p.m. |

**Victoria—Vancouver**

| | | | |
|---|---|---|---|
| Lv. Victoria | 11:00 a.m. | Ar. Vancouver | 11:45 a.m. |
| Lv. Victoria | 4:00 p.m. | Ar. Vancouver | 4:45 p.m. |

**Seattle—Vancouver**

| | | | |
|---|---|---|---|
| Lv. Seattle | 10:00 a.m. | Ar. Vancouver | 11:45 a.m. |
| Lv. Seattle | 3:00 p.m. | Ar. Vancouver | 4:45 p.m. |

**Vancouver—Seattle**

| | | | |
|---|---|---|---|
| Lv. Vancouver | 10:00 a.m. | Ar. Seattle | 11:45 a.m. |
| Lv. Vancouver | 3:00 p.m. | Ar. Seattle | 4:45 p.m. |

ONE WAY AND ROUND TRIP FARES.

| From | | To Seattle | To Victoria | To Vancouver |
|---|---|---|---|---|
| Seattle | O. W. | | 7.65 | 14.75 |
| | R. T. | | 15.00 | 28.00 |
| Victoria | O. W. | 7.65 | | 7.65 |
| | R. T. | 15.00 | | 15.00 |
| Vancouver | O. W. | 14.75 | 7.65 | |
| | R. T. | 28.00 | 15.00 | |

Above is the Victoria–Seattle and Victoria–Vancouver schedule at the start of the second year of service. Like B.C. Airways, AWA of B.C. flew a "triangle" route without a hypoteneuse, as all Vancouver–Seattle traffic had to travel via Victoria. In one year, the cost of a Victoria–Seattle ticket had plunged from $13 to $7.65.

# 11. Victoria's Elusive Air Harbour ▪ 1928–1931

From November 1928, soon after disaster struck British Columbia Airways, there had been a campaign to create a proper air harbour in Victoria. As pictures in previous chapters make all too clear, the B&K floating wharf at Shoal Point was hardly a facility to stir civic pride, especially when compared with the elegant and spacious CPR Steamship terminal on Belleville Street.

A Canadian-built Vickers Vedette from the RCAF Station at Jericho Beach is here moored at the JBAA wharf in front of the Parliament Buildings. The S.S. *Olympic* is at the Puget Sound Navigation Company dock.

Viewed at first as a modest proposal, the idea of a municipal seaplane base in Victoria's Inner Harbour stirred a controversy that extended through autumn 1931. Promoted by the City's Industrial Committee, the plan was widely supported by aviation experts, including Civil Government Air Operations in Ottawa, a branch of National Defence that relied on advice from the RCAF. Far less enthusiastic was the Dominion Department of Marine and Fisheries, not to mention steamship captains it represented: here the difference of opinion stemmed from the proposed downtown location of the base. Because of the importance of shipping for Victoria, City Council and the local Chamber of Commerce found themselves at cross purposes.

In the beginning, the question was theoretical. By the summer of 1930, however, when there was a wealthy American developer waiting eagerly to build a modern facility, the issue became far more urgent. That pending development was killed by a ruling from Ottawa. Then in 1931 there was an apparent about-face by the powerful railway and steamship companies who had been opposing a seaplane base. As cynics sourly noted, both Canadian Pacific and Canadian National had now become major stakeholders in the new Canadian Airways. Finally, there was a compromise solution that would remain in effect through World War II: from 1932 to 1943, scheduled airline departures from Victoria were available only from a seaplane base in Esquimalt Harbour.

It all began at a meeting of City Council's Industrial Committee on November 13, 1928. That body reviewed favourably a letter from the Seattle chapter of the U.S. National Aeronautic Association, suggesting

> that a municipal float for the use of seaplanes be built in the Inner Harbour. . . . Messrs. Barnes & Gorst, operating a mail service between Victoria and Seattle, would be willing to pay a nominal rental for the use of same, but of course it should be available for any seaplanes visiting Victoria.

We must remember that "Inner Harbour" in this context meant the waters between Shoal Point and Laurel Point. In the late 1920s, no one foresaw that commercial seaplane traffic would soon venture into the heart of James Bay.

Advice on these matters was then provided by senior officers from the Vancouver RCAF base at Jericho Beach. In 1928, the regular RCAF emissary to Victoria was Flight Lieutenant A.H. Hull; from 1929 to 1931, it would be Squadron Leader Earl L. MacLeod. The presence of an RCAF Vickers Vedette in front of the Parliament Buildings usually signalled a visit by one of these consultants (see photo above).

Following approved procedures, Alderman Straith's Committee learned on November 26 that a wire had come from Ottawa "advising that the Officer Commanding Jericho Beach Air Station had been instructed to come to Victoria and advise as to the best location for the municipal float for seaplanes."

An RCAF Vickers Vedette is leaving Victoria's Inner Harbour, probably in the summer of 1931. We have an exceptionally broad view to the north.

Within two days, F/L Hull had inspected a site at the corner of Erie and Dallas Road, and was willing to recommend that a licence be granted.

It was not that simple a matter, as the Victoria Harbourmaster, Captain G. Kirkendale, soon observed. On January 15, 1929, he shared with City Council's Committee his concern about log booms used for local mills. Because the lumber industry was the largest in Victoria, he preferred that the landing float be located in Esquimalt Harbour. Nevertheless the City was resolved to use three lots that it owned at the corner of Erie Street and Dallas Road, and applied to the Dominion Air Board for permission (Industrial Committee, January 25, 1929).

The Committee learned on April 16 that Captain Kirkendale and Colonel A.R. Wilby of the Marine Department preferred an adjacent site ("the Old Chemical Works property"), which would interfere less with the booming grounds. (See map and photos, pages 94–95.)

On June 13, 1929, the Air Harbour site occupied a full meeting of this Committee. S/L MacLeod had written to express his agreement with F/L Hull that all booms should be removed and kept clear. The CPR was also unhappy about the logs, it seems, since on average about twenty ships per day were entering and leaving the harbour. At this stage, nothing had yet been decided, but the outlook seemed promising.

Progress was far too slow for the Victoria Chamber of Commerce. On July 16, its Aviation Committee, headed by influential businessman and World War I pilot R.H.B. Ker, passed a motion asking City Council

> . . . to follow up closely and press for conclusion of a move made some time ago to establish a seaplane base at the foot of Erie Street.
>
> —*Colonist* 1929.07.17: 3

George I. Warren, Chamber secretary and later director of the Greater Victoria Tourist Bureau, said there had been pressure for this move from outside Victoria by "a reputable air transportation firm" (presumably Alaska–Washingon) considering an extension of its service to Victoria:

> The importance of placing the city on the air map was strongly stressed. Vancouver had spent thousands of dollars, and Seattle hundreds of thousands of dollars, in developing air terminals, and it was contended that Victoria must secure a place in aerial transportation on the continent.

Mr. Warren pointed out that City Council had taken steps some nine months earlier "to secure airport rights for an area of waterfrontage at the foot of Erie Street, but apparently no definite result had come about."

A few days later, Mr. Ker was quoted deploring Victoria's inability to accommodate properly either land-based aircraft or float planes:

> "Victoria is far behind other progressive cities in the matter of an airport," said Mr. Ker. "With the rapid strides that flying is making in every part of the world the city cannot afford to be without facilities for both airplanes and seaplanes."
>
> —*Colonist* 1929.07.23: 1

As if to underscore his warning, Alex Holden was forced to circle Victoria Harbour several times on Sunday, July 28, before he could find a landing place for the five-passenger Boeing seaplane that he was flying to Victoria. His passengers finally had to scramble over a scow near the old Hudson's Bay Wharf. Boeing publicity manager Harold Crary and George Warren, both present at the scene, were clearly dismayed (*Times* 1929.07.29.15).

With this positive nudge from the Chamber of Commerce, the City reaffirmed its request for a licence, and a news item in the *Times* on August 5 made it appear that all parties were now in agreement on the preferred site near Raymur's Point:

> The site was used formerly for mooring log rafts for local mills, and some objection to the proposal that it be devoted to a seaplane base was encountered by the city in the earlier stages of the negotiations. Now, it is understood, these differences have been ironed out, and the city awaits word from Ottawa for permission to go ahead with the plan.
>
> —*Times* 1929.08.05: 1

Though RCAF officers at Jericho Beach continued to express their unequivocal support for the plan, there were other powerful interests that strongly disapproved. Local opposition was now sparked by two highly respected mariners—Captain James W. Troup, who had retired in 1928 (at age 73) as superintendent of the CPR's Pacific Coast steamship operations, and his successor in that position, the veteran Captain Cyril D. Neroutsos (see Turner 1977: 139). Captain Troup was then chairman of the Chamber of Commerce's Harbour Committee.

September 1929, we should recall, was the heady first month of Alaska–Washington Airways' Victoria operations, based at the primitive B&K float. On September 14, 1929, the *Victoria Times* reported on page 1 that plans for the Erie Street base had been sent from Jericho Beach to Ottawa; we are told that the Department of Marine and Fisheries had already approved the concept, and the shore works and a hangar should soon be under construction. Yet ten days later, under a front-page banner headline, the same newspaper revealed that the Chamber of Commerce was now advocating a major air harbour at Esquimalt. That position had been taken after a joint session of their Harbour and Aviation committees, now combined as a Port Development Committee.

On October 1, meeting together with the City's Industrial Committee, members of this Port Development Committee presented their arguments with obvious sincerity and conviction. Captain Troup and Captain Neroutsos both spoke eloquently. Because of concern about Inner Harbour congestion, the Chamber now firmly believed that the Air Harbour must be located at Esquimalt, an alternative that both City Council and its RCAF advisors had carefully considered and rejected. Even R.H.B. Ker, who had strongly backed the Inner Harbour proposal just three months earlier, was now a convert to the Esquimalt plan. But Mayor Herbert Anscomb and Alderman William Straith were not to be dissuaded easily from their position, which they defended with repeated reference to professional advice from RCAF experts.

Telephoned on the spot, Squadron Leader MacLeod agreed to attend a similar meeting a week later, on October 8. The long and detailed minutes of that session make interesting reading, for they reveal that highly experienced mariners could disagree in 1929 with a senior professional aviator about standard procedure and protocol when ships and seaplanes share common waters. Backed by Norman Yarrow, a new face at the table, S/L MacLeod argued persuasively that Victoria's Inner Harbour was more than sufficiently spacious to allow such co-existence.

Left to themselves at the end of this stormy meeting, the Industrial Committee passed a motion asking City Council that this committee be authorized "to call for tenders for the construction of [a] seaplane base in the Inner Harbour in accordance with plans prepared by City Engineer and approved by RCAF and Department of Civil Aviation." Perhaps significantly, the Department of Marine and Fisheries is not mentioned.

In late December 1929, there was a new development that placed Victoria's Air Harbour in greater peril. Canadian National Steamships had earlier announced that it would compete with the CPR by operating its own triangle service between Victoria, Vancouver, and Seattle, with three large new vessels that would arrive the following summer—*Prince David, Prince Henry,* and *Prince Robert.* The bad news now was that these ships would be using remodelled facilities on the north (Wharf Street) side of James Bay, docks previously occupied by Pacific Salvage Company and by Evans, Coleman and Evans (*Times* 1929.12.28: 1). The Seattle-based Puget Sound Navigation Company was a long-time fixture on that side of the bay. If the Inner Harbour was already thought to be congested, it was about to get much worse.

This Canadian National Steamships triangle service was inaugurated in August 1930 by the new *Prince David.* It was destined to last for only a year, until September 15, 1931 (Turner 1977: 156). Any photographs that show a Canadian National sign above the Wharf Street docks can be dated to that brief period.

Meanwhile, the air harbour controversy was dragging on. Another acrimonious meeting was held on January 10, 1930, at the request of Col. A.R. Wilby. Backed by a letter from his Deputy Minister of Marine and Fisheries in Ottawa, he and Captain Neroutsos drew attention to current air regulations that were, they said, in conflict with S/L MacLeod's earlier assertions. These rules of the road, they claimed, "would entirely dislocate the shipping using the harbour." Again consulted by telephone, S/L MacLeod curtly dismissed their anxieties.

## The Edward A. Lowe Project

In April 1930, a potential saviour arrived on the troubled scene. Introduced to City Council in a letter from Alex Holden, he was Edward A. Lowe, Jr., a wealthy San Francisco industrialist who was president of the rather obscure Pacific International Airways. Despite the depressed global economy, Lowe had deep pockets and a desire to find new worlds to conquer. A frequent visitor to Victoria, he had a grand vision for the city's future. Apparently he saw it as the perfect hub in an aviation triangle linking Vancouver and Seattle.

Edward A. Lowe, Jr., is a shadowy figure. His proposal, considered discreetly at an Industrial Committee meeting on April 12, was to lease the City-owned Erie Street property for ten years, at the cost of a dollar a year plus taxes, and to build on it a permanent seaplane base and landing wharf. Alderman Straith's committee took immediate action to have the City Solicitor prepare the necessary agreement, to be executed by the Mayor and City Clerk on receipt of a $1,000 deposit from Mr. Lowe as a guarantee of good faith.

Further negotiations were conducted in confidence: no details appeared in local newspapers until the whole story broke on June 30. Vague hints had been made public in a front-page *Times* article on April 17, in which "former Victoria boy" Alex Holden was identified as the executive head of a new air transportation service, backed by "a group of California air-transportation men." Barely three months earlier, Holden had been a vice-president of Alaska–Washington Airways, pilot in January on a record-setting AWA flight from Seattle to Victoria. He seems now to have found a better job.

These events of 1930 might have remained mysterious were it not for a slender book called *Up in the Air*, a 1988 memoir by Mary M. Worthylake, describing her eventful fourteen-year marriage as Mrs. W.J. Barrows (1924–1938). Joseph (Joe) Barrows, a veteran California pilot, was Edward Lowe's partner in Pacific International Airways; Joe and Alex Holden hatched this Victoria scheme. After Joe and Mary Barrows had temporarily come north from California in the early months of 1930, Joe met Alex on an overnight visit to Seattle. Thoroughly familiar with all aspects of commercial aviation in the Pacific Northwest, Alex sold Joe on Victoria's unique potential, and Joe then convinced his wealthy partner that this was the golden opportunity they had both been seeking. Here is a small sample of Mary's narrative:

> During the spring and summer of 1930 Joe was on Forest Patrol work, or flying up and down the coast as far as British Columbia. He and his partner, Mr. Lowe decided to work with Alex Holden and start a triangular air service from Seattle to Victoria and Vancouver. To incorporate or operate in Canada they had to make their headquarters there, and they chose the little English city of Victoria as a base. The flights would be made in planes equipped with pontoon floats, and for a terminal they wanted landing rights to the Inner Harbor. The Canadian Pacific Railroad [*sic*] attempted to block granting of permission for planes to land there.
>
> —Worthylake 1988: 47

There were no signs of CPR opposition when Lowe's plan was first revealed on June 30, 1930. That evening's *Times* ran an upbeat story under the headline "Second Triangular Service is Announced." From his projected two-level Erie Street base, Lowe planned to operate three 11-passenger amphibians between Victoria, Vancouver, and Seattle. He was not worried about competition from Alaska–Washington Airways, for he was confident that there would be sufficient business for two lines. His ambitious plan was apparently viewed as a fait accompli, with all obstacles now removed. It was estimated that the project would initially cost $200,000 to $300,000.

For the June 30 announcement, the Edward Lowe party had flown to Victoria in two Fairchilds, piloted by Joe Barrows and Alex Holden. On July 2, those two Fairchilds took off from Lansdowne Field for a grand aerial tour of Vancouver Island; Alderman Straith, one of the ten passengers, enjoyed that day his first-ever airplane ride. The *Colonist* waxed lyrical:

> From a height of 10,000 feet off the southern edge of the Forbidden Plateau the flyers picked out Cameron Lake, Sproat Lake, Great Central Lake, the snowy peaks of the chief Island range, both East and West Coasts, and had such a bird's eye view of emerald-hued pine-girt bodies of fresh water that it was with reluctance they turned the noses of their planes away from the sight. —*Colonist* 1930.07.03: 23

Elwood White Collection

From an unrecorded source, this photo is annotated only as "Fairchilds at Lansdowne Field." These could be the two aircraft that Alex Holden and Joe Barrows flew for Edward Lowe on their 1930 Vancouver Island aerial tour.

The *Colonist* story ended with an optimistic paragraph:

> Construction on the seaplane hangar in the Inner Harbor will commence within three weeks, it is now stated, and final plans for the terminal are now being checked over.

Victoria greeted Edward Lowe as a conquering hero when he returned again on Saturday, September 20, 1930. This time he had been flown from Chicago with his wife and company director Alex Holden by pilot Joe Barrows. To herald their arrival at Lansdowne Field, Maurice McGregor landed ahead of them in Fleet CF-AKC. Both Victoria papers show Lowe being greeted by Mayor Anscomb at Lansdowne, standing in front of a new eight-passenger Consolidated Fleetster (NC-750V), one of several aircraft that he planned to equip with floats for use in the triangle service.

Pilot Joe Barrows denied rumours that the company intended to employ American pilots and workmen:

> "Although Mr. Lowe and myself are Americans, we intend to make our home in British Columbia. All Canadian men will be secured to fly the machines. The planes themselves will be registered in Canada," he said.
>
> —*Times* 1930.09.22: 20

Barrows was apparently quite serious about residing in Victoria; his wife had already rented an attractive home at Ten Mile Point. In her memoir, the former Mrs. Barrows describes arriving in the city with her two young children while Joe was still away in the east:

> When the ferry docked, the Empress Hotel and the Parliament Building grounds shone with lights. Our first view of Victoria was the most impressive I have ever had of a strange city.
>
> —Worthylake 1988: 47

On September 20, 1930, Edward A. Lowe, Jr., arrived at Lansdowne Field in this Consolidated Fleetster 17-2C (NC-750V), an 8-passenger, deluxe model priced in 1929 at $28,000. Powered by a 575hp Wright Cyclone, the plane was to be equipped with floats and put into service in Victoria. Instead, Joe Barrows and Edward Lowe moved this and their other aircraft to Alaska.

Lloyd Jarman Collection

In the livery of Seattle–Vancouver Airways, this Zenith Albatross Z-6-B, registered as NC-935Y, was owned and operated by Alex Holden between 1931 and 1934.

Thousands of Victorians crowded Lansdowne Field on Sunday, September 21, 1930 to view the visiting aircraft, which was claimed to be the fastest commercial plane now in use. In flying from Vancouver to Victoria, it had matched Gordon MacKenzie's record of 22 minutes. There was widespread local support for Lowe's project, despite mounting opposition from shipping interests. The City's Industrial Committee was still optimistic as late as October 1, but their hopes were illusory. The whole project collapsed during the last week of October. Increasingly impatient with the delay, Lowe had insisted on a firm decision. After wiring Ottawa for clarification, Mayor Anscomb was informed by National Defence on October 29 that the air harbour proposal had been refused (*Colonist* 1930.10.31: 1).

Discouraged by this turn of events, Lowe and Barrows decided to move their Pacific International Airways to Alaska. While en route, they became engaged in the widespread search for lost Canadian pilot E.J.A. (Paddy) Burke, a search already compounded by the disappearance of Alaska–Washington Vega NC-103W, flown by Canadian pilot Robin (Pat) Renahan (page 96). Their adventures are related by Mary Worthylake, whose husband Joe Barrows was involved in a serious double crash at Telegraph Creek, where the new Fleetster fell through the ice (see Corley-Smith 1993 and Schofield 2004).

Later in 1930, Alex Holden joined Joe Barrows as a pilot for Pacific International Airways in Anchorage.

Alex and his wife Lorraine are often mentioned in Mary Worthylake's book. The Holdens moved back to Seattle in February 1931, for the birth of their son Alex. There for three tough Depression years Alex Sr. eked out a living with his Zenith biplane (photo above). In 1934 he returned to Alaska, becoming successful and famous in the southeastern Panhandle. Bush pilot and airline founder (Marine Airways, Alaska Coastal Airlines), he enjoyed a remarkable career.

In the early 1940s, Alex Holden Jr. attended Victoria's Glenlyon School.

## Canadian Airways Boldly Arrives in Victoria's Inner Harbour

The imposing figure in the light grey suit (centre left) is Victoria Mayor Herbert Anscomb. The date is June 30, 1931, and he is celebrating the arrival of Canadian Airways' special inaugural flight between Vancouver and Victoria, with regular service scheduled to begin the next day.

Mayor Anscomb may still be in shock at the fact that this new service has begun with no civic effort on his part, after he and his Council had struggled vainly for almost three years to develop a modern air terminal near Shoal Point. And the almighty Canadian Airways is allowed to bring its commercial flights directly to this JBAA wharf in front of the B.C. Legislature! It must be gratifying to have such influence in Ottawa.

At the controls on this flight was Major Donald R. McLaren, D.S.O., M.C. and bar, D.F.C., etc. (far right). Canadian Airways' Vancouver manager, he would be followed in July by less exalted pilots on the CAL roster. This inaugural group includes Harold Husband, Canadian Airways' Victoria traffic manager (far left); Vancouver Mayor Louis D. Taylor (who was almost decapitated at a similar event in 1928); R.J. Cromie of the *Vancouver Sun* and his young son; Charles Webster of the Vancouver Publicity Bureau; and representatives of the Victoria Chamber of Commerce and Vancouver Board of Trade.

Moored at the wharf is one of four Boeing C-204 Thunderbird flying boats manufactured in Vancouver by Boeing Canada. Powered by 410hp P&W pusher engines, they normally accommodated five passengers. Canadian Airways flew CF-ALA, CF-ALB, and CF-ALC.

Western Canada Airways, Ltd. had evolved into Canadian Airways, Ltd. The crest had evolved likewise. There are two officially authorized and thoroughly reliable histories of the Company, the second of which extends its scope to cover the development of Canadian Pacific Air Lines:

Molson, K.M. 1974. *Pioneering in Canadian Air Transport.* Altona, Manitoba: D.W. Friesen.

Bain, Donald M. 1987. *Canadian Pacific Air Lines: Its History and Aircraft.* Calgary: Kishorn Publications.

Because the Board of Canadian Airways had equal representation from Canadian Pacific and Canadian National Railways, under the strong chairmanship of Winnipeg's James A. Richardson, it was viewed informally as Canada's National-Airline-in-Waiting, and for that reason it was thought to be almost omnipotent. Still, even Canadian Airways was required after the summer of 1931 to accept Ottawa's verdict that Victoria's Inner Harbour was not large enough to accommodate both shipping and seaplanes. Only for the months of July to September 1931 were scheduled flights allowed to venture into James Bay; then the Inner Harbour ban on commercial aviation was reaffirmed.

In June 1931, D.B. Plunkett, M.P. for Victoria, persuaded the Department of National Defence to relax its opposition to air traffic into Esqumalt Harbour. That would be Victoria's seaplane base for the 1930s.

The three photographs to the right were all taken in the summer of 1931. The top picture shows a Canadian Airways Boeing C-204 Thunderbird, most likely CF-ALA, at the JBAA wharf, with the new Imperial Oil Causeway Garage and tower in the background. The lower two photos can be dated to Saturday, August 1, 1931, when Canadian Airways inaugurated its Victoria–Seattle service. The middle picture shows CF-ALA arriving at 10:30 a.m. with Seattle Mayor Robert Harlin aboard. The grainy group photo ran in the *Times* on August 5. It shows Mayor Harlin with Victoria dignitaries who are about to enjoy the return trip to Seattle. The pilot on the incoming flight was E.C.W. (Ted) Dobbin.

As was apparent with both B.C. Airways and Alaska–Washingon Airways, Seattle and Victoria then enjoyed a very close relationship.

L–R: George I. Warren (Victoria & Island Publicity Commissioner); Harold Husband (Canadian Airways' Victoria Manager); Robert W. Mayhew (President, Victoria Chamber of Commerce); L.S. McIntyre (Transportation Division, Seattle Chamber of Commerce); Seattle Mayor Robert Harlin; Victoria Mayor Herbert Anscomb; and pilot E.C.W. (Ted) Dobbin.

## Coastal Airways: Possible Competition for Canadian Airways?

Midsummer 1931 witnessed the appearance of a new passenger air service between Victoria and Vancouver: Coastal Airways B.C. Ltd., known also (confusingly) as Royal Air Lines. Presenting itself as successor to Alaska–Washington Airways B.C., it had some impressive backers: Vancouver's Richard J. Gosse of B.C. Packers, Victoria's Robert P. Butchart (President), and Squadron Leader J.H. "Tuddy" Tudhope, former C.O. of Jericho Beach RCAF Station. The story of its brief and controversial existence has been explored by Peter Corley-Smith (1993: 112–114).

Because of practical inexperience and internal dissent, Coastal Airways *alias* Royal Air Lines had a very short life. It acquired a Bellanca Pacemaker (CF-AND), which had been part of the recently concluded Trans-Canada Air Pageant; this was to fly the route along with a Fairchild 71, probably the familiar CF-AJP. The Bellanca appears in the three pictures to the right, at least two of which date from August 1, 1931. In the bottom photo—a striking view of the Causeway and Inner Harbour—the nearer of the two aircraft is a Boeing C204 Thunderbird of Canadian Airways, Royal Air Lines' arch-competitor.

In the top picture, Pilot C.E. (Pat) Kelly stands on the Pacemaker's right float, before the colourful background of Victoria's Inner Harbour. He had been recruited to fly as backup to S/L Tudhope. Pat Kelly appears third from the right in the middle photo, which shows Victoria Postmaster G.H. Gardiner and other officials gathered to mark Royal Air Lines' inaugural airmail flight to Vancouver on Saturday, August 1, 1931. This achievement was highly acclaimed in the press: some 25,000 letters crossed the Strait that day.

TRAVEL the Triangle Route in comfort and safety the modern, convenient way. The luxurious, dependable Fairchild 71 and Bellanca Pacemaker planes of the Royal Air Lines take you swiftly to Vancouver and Seattle from their base in the heart of the city . . . just below the Empress Hotel.

*Schedule*

*Effective July 30*

Lv. Vancouver - - - - 9.00 a.m.
Ar. VICTORIA - - - 9.50 a.m.
Lv. VICTORIA - - - 10.00 a.m.
Ar. Seattle - - - - - - 10.50 a.m.

Lv. Seattle - - - - - - 11.15 a.m.
Ar. VICTORIA - - - 12.05 p.m.
Lv. VICTORIA - - - 12.15 p.m.
Ar. Vancouver - - - 1.05 p.m.

Lv. Vancouver - - - - 2.00 p.m.
Ar. VICTORIA - - - - 2.50 p.m.
Lv. VICTORIA - - - - 4.30 p.m.
Ar. Vancouver - - - 5.20 p.m.

*Fares*

Victoria-Vancouver - - - - - - - - - - Single, $7.00 Return, $12.00
Victoria-Seattle - - - - - - - - - - - - Single, $7.00 Return, $12.00
Vancouver-Seattle - - - - - - - - - - - Single, $12.00 Return, $20.00

*Tickets and reservations from*
C & C TICKET OFFICE
906 Government St. Phone Empire 1121

*"Fly with His Majesty's Mail"*

VIC-1

ROYAL AIR LINES

OPERATED BY COASTAL AIRWAYS B.C. LTD.

Canadian Airways ran a huge display ad in the *Victoria Times* (above left) on June 29, 1931, announcing a service to start on Dominion Day. Royal Air Lines (Coastal Airways) launched a parallel service one month later with its own flashy ad—offering cheaper fares. Lacking resources and experience, the upstart Royal Air Lines collapsed within days, despite having been selected to carry mail between Victoria and Vancouver. At right is Royal's inaugural flight on August 1 from Vancouver to Victoria. For the next decade, Canadian Airways was the only company to hold Victoria–Vancouver airmail contracts.

August 1, 1931: Pat Kelly and postal officials with Royal Air Lines' Bellanca.

## Coast Air Transport and Wells Air Transport: More Competition?

Two airlines that played a significant role during the early 1930s were Coast Air Transport, headed by Adam Richardson, and Wells Air Transport, headed by W. Hunter Wells. Both were based in Vancouver, but briefly offered flights to and from Victoria. (Coast Air Transport had nothing to do with Coastal Airways!)

It is a curious fact that the upper two pictures on this page must date from the summer of 1932, when seaplane traffic was theoretically banned from Victoria's Inner Harbour. Apparently exceptions could be made for inaugural flights or important promotional photo opportunities.

In July 1932, Adam Richardson bought Boeing C-204 flying boat CF-ALA, earlier leased to Canadian Airways, and began to compete with that airline for Victoria passengers. Flights between Vancouver's Sea Island and Esquimalt Harbour were probably limited to charter and special services; but Coast Air Transport seems to have begun a scheduled service between Victoria and Seattle. There is little evidence left today beyond the 1932 photo to the right, in which Victoria Mayor David Leeming presides over another Victoria–Seattle inaugural ceremony.

On November 19, 1932, Wells Air Transport inaugurated a winter service between Esquimalt and Wells Air Harbour, Sea Island, exploiting the fact that Canadian Airways was not yet offering year-round flights on this route. Alaska–Washington pilot Gordon MacKenzie was back flying his old aircraft, Fairchild CF-AJP. Heading a large civic delegation, Mayor David Leeming welcomed Vancouver RFC veteran and company owner William Hunter Wells (*Colonist* 1932.11.20: 20).

Boeing C-204 Thunderbird CF-ALA in 1932, now operated by Coast Air Transport. Left, owner Adam Richardson; right, pilot E.C.W (Ted) Dobbin.

B.C. Archives G-00334

With CF-ALA sporting new livery, Coast Air Transport launches its 1932 service to Seattle. Mayor David Leeming is present, flanked (left) by pilot Ted Dobbin and owner Adam Richardson, with Victoria businessmen Harold Husband, R.H.B. Ker, Herbert Stevens, and James H. Beatty.

Here we see Fairchild 71 CF-AJP at Cordova Bay, likely a charter flight during the Wells Air Transport period. McMorran's Pavillion is far right.

# 12. Esquimalt Serves as Victoria's Air Harbour ▪ 1932–1945

Esquimalt did not become a regular air harbour until May 1932, when Canadian Airways resumed its Victoria–Vancouver service after a winter hiatus. This Esquimalt to Sea Island route was inaugurated by the Sikorsky amphibian CF-ASO, ferried west to B.C. that spring by pilot Ted Dobbin (above, second from left). For the next ten years, scheduled flights in and out of Victoria were based at one of two ramps near Signal Hill, either at the foot of Canteen Street, close to Yarrows (photos on this page) or at the foot of Pioneer Street, beside End House Pub. Flying boats and seaplanes were the only regular means of air transport from Victoria. Commercial flights did not use James Bay again until long after World War II.

Sikorsky amphibian S-38-C (CF-ASO) is seen against the backdrop of Signal Hill. It was powered by two 420hp P&W Wasp engines. This was the primary aircraft on the Victoria–Vancouver route from May 12, 1932 until May 9, 1934, when it had to be withdrawn because of severe salt-water metal corrosion. In 1932, the schedule called for two round-trip flights a day, leaving Sea Island at 10 a.m. and 4 p.m., and Esquimalt at 11:15 a.m. and 5:15 p.m.

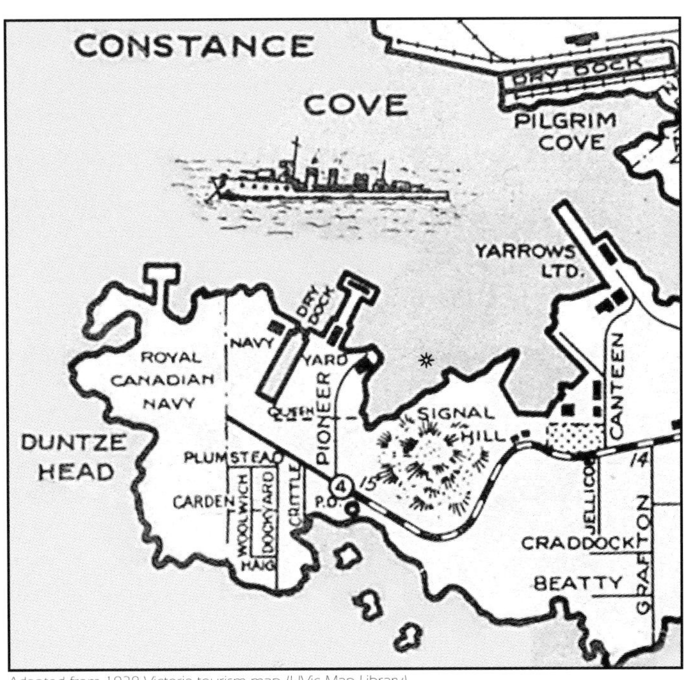


Adapted from 1938 Victoria tourism map (UVic Map Library)

(Above and across the page, above):
Esquimalt Air Harbour, summer 1933. In the photo above, the large white building to the left is The End House, at the bottom of Pioneer Street. The twin-engine Sikorsky flying boat can be seen taxiing out for takeoff, at the point marked by a star on the map (left). Four other planes remain moored in the harbour, but are very hard to distinguish. In the picture opposite, the Sikorsky is gone, and the other four may appear somewhat more clearly: they are Boeing C-204 flying boat (CF-ALA), now Coast Air Transport, Vancouver; Fleet 2 (CF-ANF), owned by A.H. (Hal) Wilson and W.G. Mackenzie, Vancouver; the former Alaska–Washington Fairchild 71 (CF-AJP), now Wells Air Transport; and Fokker F-11-AHB (CF-AUV), owned by Bill Strong of Tulsequah, B.C. It is a surprising number of aircraft. The Canadian Airways Sikorsky carried most of the traffic between Victoria and Vancouver, but the other four planes were all engaged in some form of commercial aviation.

(Below) The Sikorsky Amphibian could operate on wheels at Vancouver's Sea Island Airport.

Maurice McGregor Collection

Canadian Airways pilot William J. (Bill) Holland, who appears in the two photos to the right, was one of the pilots who flew the Sikorsky amphibian flying boat in the summer of 1932. In the upper picture, he is posing with his flight engineer. The landing wheels are visibly retracted, in contrast to the lower photo, which was taken at Sea Island.

Bill Holland was a skilled bush pilot and Canadian Airways veteran who spent some time with the RAF Atlantic Ferry Command during World War II. Not long after returning to civil aviation with Canadian Pacific Airlines, he was killed while flying co-pilot with captain Ernest Kubicek on a CPAL flight from Prince George to Vancouver. On December 20, 1942, their Lockheed 14 (CF-CPD) crashed into Mount William Knight in the Cheam Range, near Chilliwack, killing all thirteen aboard. At that time, it was the worst crash in the history of B.C. aviation.

The *Canadian Airways Bulletin* is a superb source of official information. In addition to schedules and informative articles, it provides regular lists of the aircraft, pilots, and mechanics stationed at its various bases across the country (see below, right).

Elwood White Collection

*CAL Bulletin*, December 1934

From 1934 to 1936, Fokker Super Universal G-CASQ often carried both mail and passengers on the Victoria run. Above, it is shown at Port Neville, on Vancouver Island south of Alert Bay.

| VANCOUVER BASE | | | | |
|---|---|---|---|---|
| Vancouver | CF-AKY | Fairchild F71 | MacLaren | Haslett |
| | G-CASQ | Fokker Super | Wells | Reid |
| | CF-ALA | Boeing BIE | Lawson | Terry |
| | CF-ARF | Boeing Totem | McGregor | Bulger |

*CAL Bulletin*, August 1, 1935

Pilot E.P.H. (Billy) Wells. Wells and F. Maurice McGregor were perennial "frequent flyers" on Canadian Airways' Victoria–Vancouver route.
*CAL Bulletin*, December 1933

*CAL Bulletin*, December 1934

*CAL Bulletin*, December 1933

Before he was assigned to passenger flights, Maurice McGregor (above) flew on fisheries patrol. In Fokker Super Universal G-CASQ, with Fisheries Supervisor J.F. Tait (left), he checks a reported violation at French Creek .

In 1934 and 1935, Canadian Airways flew two aircraft on the Vancouver–Victoria route: Fokker Super Universal G-CASQ and Fairchild 71 CF-AKY, shown above moored in Esquimalt, at the foot of Pioneer Street. Here we see the relationship between the seaplane float and The End House inn and beer parlour, thought by many to be the perfect airport waiting room. It was the official Canadian Airways terminal until 1942.

Although he compiled an outstanding lifetime record of air safety, Victoria's Maurice McGregor had one serious mishap with Canadian Airways' Fairchild 71 (CF-AKY), in which luckily no one was seriously hurt. It was not a regular Victoria–Vancouver flight, but a special charter for mining executives on May 16, 1935. After landing in Vancouver Harbour, Maurice was taxiing towards shore when he inadvertently ran into a tow line of the CPR tug "Nanoose," causing the Fairchild to flip over and toss its passengers into the water (see photograph, right). No blame was attached to the accident, and McGregor was commended for rescuing one of his passengers from possible drowning.

Maurice McGregor Collection

Tom Brundson Collection

In 1935 and 1936, Boeing C-204 Flying Boat CF-ALA (left) returned to service on Canadian Airways' Victoria–Vancouver run. The seaplane behind it is a de Havilland Moth, quite likely Terry Finney's CF-AGM (see page 113).

**VANCOUVER—VICTORIA**

Vancouver, Ground Floor, Marine Building..........Phone Seymour 6401
Vancouver—for Reservations.......... " Seymour 7131
Victoria Agent.......... " Empire 0222

PASSENGERS, MAIL, EXPRESS—DAILY EXCEPT SUNDAY

9.30 a.m., 4.00 p.m. .. Lv.......Vancouver.......Ar..... 11.15 a.m., 5.45 p.m.
10.15 a.m., 4.45 p.m. .. Ar.......Victoria..........Lv..... 10.30 a.m., 5.00 p.m.

**Air cruises, sightseeing tours, fishing and hunting trips arranged for at special rates.**

CAL Bulletin, August 1935

In midsummer 1934 and 1935, Canadian Airways ran two daily round-trip flights between Victoria and Vancouver, but aircraft were scheduled to return each night to hangars and other facilities on Sea Island.

An RCAF Vickers Vancouver is moored at the Pioneer Street terminal in Esquimalt Harbour during the mid-1930s. Although RCAF flying boats from Jericho Beach had been using Esquimalt Harbour since 1920, this wharf and terminal were developed by Canadian Airways, whose aircraft were its principal users from 1932 to 1942. In this photograph, we are looking northeast from the Dockyard and Signal Hill to Yarrows Shipyard (right) and the Esquimalt Drydock (left).

Canadian Airways would not celebrate 1937 as one of its better years. After introducing spiffy uniforms and buying two Lockheed Electra aircraft in order to play a key role in Canada's new national airline, James Richardson saw his plans unravel: many of his key employees transferred in the summer of 1937 to a Trans-Canada Air Lines that totally excluded Canadian Airways (see page 116).

Captain Billy Wells and First Office Maurice McGregor, still dressed in their new Canadian Airways uniforms, flew the inaugural TCA flight from Vancouver to Seattle on September 1, 1937, aboard Lockheed Electra CF-AZY. Meanwhile, Victoria–Vancouver service had been suspended for 1937, and would not resume until March 1939.

All seaplane flights were long since banned from Victoria's Inner Harbour—even the seasoned veteran Percy Barnes had to fly the Seattle air mail into Esquimalt. Otherwise, there was very little commercial aviation anywhere in Greater Victoria until 1939. In the spring and summer of 1938, W.J. Dyson and Terence H. Finney made an effort to satisfy Victoria's various needs with three aircraft based at Esquimalt, but this enterprise had only limited and temporary success (see next page).

Canadian Airways returned in March 1939 to serve Victoria, followed by Grant McConachie's new Canadian Pacific Air Lines, which absorbed Canadian Airways in 1942. By 1943 there were five round trips daily.

Maurice McGregor, 1937. From Canadian Airways to TCA.

Maurice McGregor Collection

Maurice McGregor Collection

Boeing Totem CF-ARF was a one-of-a kind aircraft, a four-place pusher monoplane flying boat manufactured in 1932 by Boeing Aircraft of Canada, Vancouver. Maurice McGregor (in both photos above) flew it often for Canadian Airways in the mid-1930s—regularly on fishery patrol or charter flights, and occasionally on the regularly scheduled service between Esquimalt Harbour and Vancouver Airport.

Victoria Is Back on Air Lanes Once More

Financed by a small group of local businessmen, Victoria is to have a regular air service with the Mainland again. Last evening the first scheduled round trip was flown; while earlier in the afternoon a special flight from Esquimalt to Sea Island was undertaken, with Pilot Terence Finney in charge, the passengers including Fred M. McGregor, Alderman Lloyd Morgan and C. S. Henley. The photograph was taken at Esquimalt just before the takeoff. Left to right: W. J. Dyson, F. M. McGregor, Alderman Morgan, C. S. Henley and Walter S. Miles. Pilot Finney is seen in the open port of the Waco plane, which arrived here Tuesday from the factory.

*Colonist* 1938.05.19

Hoping to re-establish a Victoria–Vancouver flying service, W.J. Dyson bought in May 1938 a brand new four-passenger Waco YKC-2 seaplane with red fuselage and silver wings (CF-AWL, left), entering partnership with veteran pilot Terence H. Finney, who already owned a DH 60M Moth (CF-AGM, below). This short-lived enterprise, called Island Airways, was backed by a group that included Maurice McGregor's air-minded father, insurance executive Frederick M. McGregor, seen in the photo, left.

The Waco arrived May 17, 1938, and began twice daily Victoria–Vancouver service the very next day—upsetting Air Inspector R. Carter Guest, who had not yet granted his approval.

Earlier in 1938, Dyson had bought Canadian Airways' Boeing Totem CF-ARF (p.112), a unique flying boat previously flown often by Maurice McGregor, now a TCA pilot.

On October 3, 1938, the flying season for the Dyson-Finney partnership came to an abrupt end with a spectacular crash of DH Moth CF-AGM in Esquimalt's Plumper Bay, near the Songhees Reserve. This seaplane is shown right, in happier days, at Vancouver's Sea Island Airport. It was considered miraculous that the two occupants survived an almost vertical dive into a log boom, from a height of about 200 feet. The aircraft was completely demolished. Terry Finney's injuries were more serious than originally supposed, but his student Ralph Bonner suffered little more than a broken nose. Finney had been pulled from the plane by Alfred Nix, a boom man working nearby. Bonner was a prominent Victoria athlete, who was receiving his second flying lesson. The incident was widely publicized in British Columbia newspapers.

Terence H. Finney was a Royal Air Force veteran who had emigrated to Canada in 1929. He tried twice to establish a flying school at Esquimalt, in 1934 and 1938, but was much more closely associated with aviation in Vancouver, where he became secretary-manager of the Aero Club of British Columbia. Two of his well-known Victoria student pilots were Bill Sylvester (Chapter 18) and Marjorie Todd, who was photographed alongside CF-AGM in May 1938 (*Victoria Times*: 1938.05.21: 7).

The British Columbia Archives' vital statistics website reveals that Terence Hollingdale Finney died in Haney, B.C., on January 16, 1978, at the age of 75.

**NO. 249—EIGHTIETH YEAR**

**CHEAT DEATH WHEN PLANE CRASHES ON BOOM AT ESQUIMALT**

**Captain "Terry" Finney and Ralph Bonner, Pupil, Escape Serious Injury When Seaplane Dives Out of Control—Engine Failure Believed Cause—Machine Reduced to Matchwood**

*Colonist* 1938.10.04: 1

For the two-year period 1937 and 1938, there were no Canadian Airways flights between Victoria and Vancouver, but regular service was resumed on March 1, 1939. Because the new Trans-Canada Airlines had no airport facilities in Victoria, Canadian Airways was again awarded the airmail contract. During the month of March 1939, Canadian Airways Fairchild 71 (CF-BKP) made one round trip daily between Esquimalt Harbour (right) and Sea Island (middle photo right), transporting mail and passengers. Its schedule was arranged so as to provide connection with the national TCA service to eastern Canada.

At lower right is an excellent picture of the End House Pub, surrounded by an unusual number of automobiles. We are looking down Pioneer Street towards the Air Harbour, with Yarrows Shipyard visible in the distance. This photograph was probably taken on March 1, 1939, the day when Canadian Airways resumed its Victoria–Vancouver passenger and airmail service with Fairchild CF-BKP.

Tom Brundson Collection

Gordon Ballentine Collection

Captain Gordon Ballentine, 1940. In a letter to Elwood White (March 9, 1971), this veteran Vancouver pilot—seen here with a DH-89 Rapide—estimated that he had completed 1,800 flights between Victoria and Vancouver. He began in June 1939 with a Bellanca Pacemaker (CF-BFC). His log book listed six other aircraft on this route: Fairchilds CF-BKP and CF-AWV; and four DH-89 Rapides, CF-AYE, CF-BBH, CF-BND, and CF-BNG.

Tom Brundson Collection

Rapide CF-BBH was named City of Victoria. It appears above and left, with Captain Walter Gilbert (see p. 76); note the Canada Post Office crest. Below is CF-AYE (*Zeballos Express*), a Sea-Island-to-Esquimalt regular.

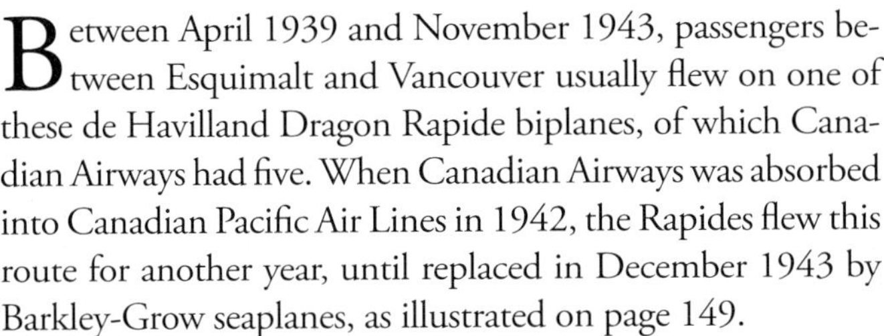

Between April 1939 and November 1943, passengers between Esquimalt and Vancouver usually flew on one of these de Havilland Dragon Rapide biplanes, of which Canadian Airways had five. When Canadian Airways was absorbed into Canadian Pacific Air Lines in 1942, the Rapides flew this route for another year, until replaced in December 1943 by Barkley-Grow seaplanes, as illustrated on page 149.

In March 1942, the original Esquimalt Air Harbour was taken over by HMCS Dockyard, and the seaplane base was moved to View Royal's Limekiln Cove. During these war years, Western Air Command required passenger planes to keep their window blinds closed during takeoff and landing. This was true also of TCA planes at Patricia Bay during the last three years of World War II.

It was unusual for two Canadian Airways Rapides (here CF-BBH and CF-BND) to be moored together at Esquimalt. The occasion was likely August 1, 1940, as British film star Anna Neagle arrived by Canadian Airways seaplane CF-BND to perform with Gracie Fields in a special Victoria benefit show for the Navy League (*Times* 1940.08.01: 1).

# 13. The Difficult Birth of Trans-Canada Air Lines ▪ 1937

In the year 1936, Canadian Airways worked closely with the Federal Government in planning a truly national airline. Because its own creation in 1930 had involved both Canadian Pacific and Canadian National Railways, and it was already viewed as the *de facto* national airline, Canadian Airways expected to play a dominant role in the new public venture. In 1936 it developed its new Vancouver–Seattle run as a trial operation for the imminent national service. Buying two new Lockheed 10A Electras, CAL installed new radio equipment and heating systems, and outfitted pilots Billy Wells and Maurice McGregor in smart new dark blue uniforms. Sadly for the Company, it did not become part of Trans-Canada Airlines.

The story has been told many times (see Molson 1974 or Pigott 2001). The problem was partly a disagreement over the structure and control of the new Board of Governors, in which the CPR felt outnumbered, and partly other disputes between that Company and the Liberal Government's chief negotiator, the powerful C.D. Howe.

On April 10, 1937, Parliament passed the Trans-Canada Air Lines Act; TCA operations began September 1, 1937, with those same two Lockheed Electras on the Vancouver–Seattle run. The captain and first officer of Electra CF-AZY were Billy Wells and Maurice McGregor, still wearing their Canadian Airways uniforms. Along with twenty other pilots and senior executive Donald R. MacLaren, they had been released to join TCA.

One consequence was the withdrawal of Canadian Airways' Victoria–Vancouver service between October 31, 1936 and March 1, 1939. To make matters worse, it appeared that Victoria's lack of an airport might exclude it entirely from the future plans of Trans-Canada Airlines, a prospect that caused great concern in the B.C. Capital. This was a dismally low point in local aviation history.

By the time Canadian Airways service was restored in March 1939 (see page 114), the prospect looked much brighter, and Victorians became keen supporters of the new Trans-Canada Air Lines. On March 15, 1939, Tom Merriman of the *Victoria Times* and Hugh MacCallum of the *Colonist* flew from Esquimalt Harbour to join the inaugural TCA overnight flight from Vancouver to Montreal, on Lockheed 10A Electra CF-TCA. Merriman and MacCallum filed graphic stories of this historic trip.

March 15, 1939. A group of Victoria newsmen wait near Esquimalt's End House for the departure of a Canadian Airways seaplane that will fly Tom Merriman (the short man centre left) and Hugh MacCallum to Vancouver, where they will join the inaugural TCA overnight flight to Montreal by Lockheed Electra. The plane to Vancouver, flown by Norville Everett (Molly) Small departed at 5:30 p.m. Their reports appeared in Victoria papers later that month, and the event was recalled at length by *Times* City Editor Archie Wills in the *Colonist Islander*, 1958.08.27:12–13.

# PART FIVE

# PATRICIA BAY RCAF STATION

## 1936–1945

DND Photo courtesy *Victoria Times Colonist*

In clockwise order, from upper right: Bristol Bolingbroke, Lockheed Hudson, North American Harvard, Stinson 105, Noorduyn Norseman, Supermarine Stranraer, Blackburn Shark, Northrop Delta, Avro Anson, Bristol Beaufort, Westland Lysander.

# 14. The Royal Canadian Air Force at Patricia Bay ▪ 1936–1945

In the archival records of the Department of National Defence (see bibliographical note below), the origins of Patricia Bay Airport can be traced back to September 1936. Even before any feasibility study was undertaken, word had seeped out that there was interest in a military air base on Vancouver Island, and enquiries began to arrive in Ottawa from real estate agents and property owners as far apart as Sidney and Long Beach. On September 11, 1936, Canada's Controller of Civil Aviation, J.A. Wilson, stated in an internal memorandum that his branch was not contemplating an aerodrome on Vancouver Island, but he noted that there was renewed interest in the establishment of an airport at Victoria, and that the Sidney Board of Trade had been active in connection with a site in that vicinity.

A full three years before the outbreak of World War II, the military interest was abundantly clear. In responding to the same internal enquiry that had been addressed to Mr. Wilson, Air Commodore George M. Croil, Senior Air Officer of the RCAF, confirmed that he had just recently taken exploratory action:

> I have written confidentially to the Officer Commanding No. 4 (F.B. [= Flying Boat]) Squadron, Vancouver, asking him to search for and report upon suitable sites for aerodromes at or near Victoria, and at or near Alert Bay. This letter of mine has not yet reached the O.C., No. 4 (F.B.) Squadron. Patricia Bay, near Victoria, has been suggested for investigation (921-1-28, September 18, 1936).

On October 29, 1936, a confidential report was submitted by Wing Commander A.B. Shearer, C.O. of this No. 4 (Flying Boat) Squadron at Jericho Beach:

> Subject: *Possible Combined Landplane & Seaplane Site, Patricia Bay, Vancouver Island*
>
> 1. It is desired to recommend for R.C.A.F. Headquarters consideration a possible site at Patricia Bay, Vancouver Island for a combined seaplane and landplane station.
> 2. Two flights were made by the undersigned over the area in the vicinity of Patricia Bay and Sydney [*sic*]. I am convinced that this site . . . is the only suitable landplane and seaplane site available at the Southern end of Vancouver Island.
> 3. It is suggested that if a more detailed report is desired on this property it could be obtained through Headquarters M.D. XI. It is considered desirable, from an engineering point of view to know if it is practical to build a seaplane hangar on piles at the end of the small bay immediately East of the old wharf . . . .
> 4. The other alternative would be to use the present embankment and wharf as an apron and bring the aircraft across the road to a hangar built on the Southeast side. Aircraft could be launched from the wharf with a crane, if desired.
> 5. The rails have been removed from the line marked Canadian Northern Railway on the map. The approach to the dock and the dock itself have not been in use for some years.
> 6. No difficulty is anticipated in developing a good service aerodrome on site No. 1.

This site No. 1 must be the more-or-less level terrain that we now know as Victoria International Airport, which was not quite so level in 1936. When we pass through that modern facility in the 21st century, we may not realize that a major factor in its selection was the potential for an adjacent seaplane base. The seaplane facility at Patricia Bay is still very much in use, but its relative importance has dwindled since World War II.

On December 15, 1936, the *Victoria Times* was more than a little premature in its choice of a front-page headline: "PATRICIA BAY BECOMES BASE." As the text of this news report indicates, the so-called base was merely a flurry of exploratory activity, but it was stated as public knowledge that two flying boats from Jericho Beach would take up temporary headquarters at Pat Bay, with the likelihood of a machine shop and a service depot to follow. The *Times* does add one telling piece of information:

> A number of summer cottages, owned by Victorians, have been rented by mechanics and aviators and others from Jericho.

The source of factual information on pages 118 to 120 is a 33-page chronological summary compiled within the Department of National Defence from official DND correspondence and the Station Diary of the Commanding Officer at Patricia Bay. A copy of this chronicle was mailed on request to Elwood White around 1970, for his use in preparing a history of aviation in Victoria. The authors of this book have not been able to consult archival documents in Ottawa, and therefore make no claim to original research on the Patricia Bay Station. We are indebted to aviation historians for their careful research on individual RCAF squadrons, in particular the comprehensive study of Kostenuk and Griffin (1977) and Carl Vincent's articles (1980) on 111 (CAC) and 111 (F).

The DND chronicle describes that first on-site reconnaissance at Pat Bay. W/C A.L. Cuffe was now C.O. of No. 4 (FB) Squadron at Jericho.

> In Dec., 1936, W/C W.R. Kenny, W/C Cuffe, S/L L.F. Stevenson and S/L E.L. MacLeod investigated this site. They reported that it was "the best looking potential aerodrome around Sidney or Victoria and possibly in the southern portion of the island. Would make a very good service aerodrome." At the same time a report was presented on the potential merits of Patricia Bay as a seaplane base. "Patricia Bay appears to be the best seaplane operating base on the south end of Vancouver Island [or] the lower mainland. Its proximity to the naval base at Esquimalt and District Headquarters, Victoria is an added advantage. An aerodrome (at Sidney) could be developed within one mile of the bay."

On February 5, 1937, Air Commodore Croil wrote a confidential memo from Ottawa to Wing Commander Cuffe. He has obviously been persuaded by the reports from Vancouver, but requests further information from Meteorological Service observers in Vancouver and Victoria. By April 1, 1937, W/C Cuffe is able to submit a detailed and confidential recommendation to the Department of National Defence. This very interesting report (File 6651) runs to four legal-sized pages, single spaced, and is a thorough summary not only of the advantages and possible limitations of Patricia Bay, but also of the broad strategic and tactical assumptions that would underly the role of Western Air Command in World War II. Not all of W/C Cuffe's predictions were clairvoyant, but we can watch the wartime aerial defensive plan for Western Canada already taking shape.

In April–May of 1937, when the official decision on Patricia Bay Airport was still almost a year away, there was a timely exchange of correspondence between the Deputy Ministers of Transport and National Defence on the need to coordinate civil and military aviation. The year 1937 would see the creation of the long anticipated Trans-Canada Air Lines, and there was a very reasonable concern that the Department of Transport (DOT) should be kept fully informed about any possible development by the RCAF of airport facilities on southern Vancouver Island. As events transpired, Victoria and Patricia Bay would miss out on the first round of TCA service, and not until June 1943 was Vancouver Island fully linked to the national airline. As we shall see, there were other factors involved in that six-year delay.

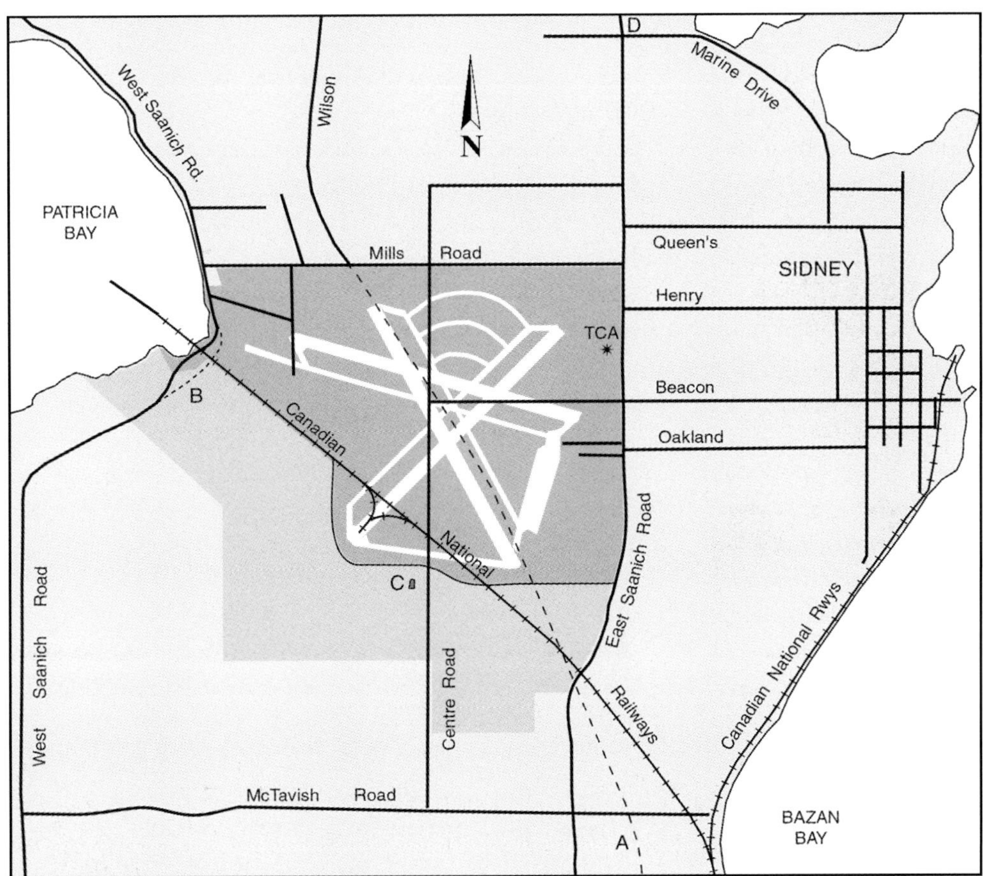


SIDNEY AND NORTH SAANICH, 1935–1936
(with the first stage of the future Patricia Bay Airport superimposed)

The CNR tracks to Patricia Bay were lifted in 1935, leaving a firm gravel road across the airport site.

Broken line "A" traces the former route of the long-abandoned B.C. Electric Interurban Railway, which used to continue along Wilson and Tatlow to Deep Cove. Dotted line "B" shows a minor future relocation of West Saanich Road. The structure labelled "C" was the North Saanich Consolidated High School, built in 1936, but purchased by DND in 1942, with provision for a new school "D" north of Sidney.

The future airport runways appear in white; even after two expansions of the main runway, their original position has never changed.

The mid-grey shading shows land acquired for the airport in Stage 1, but not used during World War II.

For all practical purposes, the Patricia Bay decision had been made by June 1937, when a site survey was commissioned and a budget prepared for land acquisition. If one must identify a date for the creation of the Airport, the best choice might be March 7, 1938, when the Committee of the Privy Council concurred in a recommendation to acquire the property (P.C. 427). A detailed proposal had been submitted to the Governor General in Council by C.G. Power, Acting Minister of National Defence, on February 26, 1938. This may be regarded as the Patricia Bay Charter, since it precisely defines the boundaries of the site, then comprising about 700 acres, with land and improvements currently assessed at $162,430:

> "It is estimated that options for purchase of the above properties can be secured for the sum of $210,000. This price is considered reasonable in view of assessed values being generally much below market values."

Mr. Power added that most of the needed amount was available in the Vote for the Royal Canadian Air Force 1937–38. In effect, Canadian taxpayers had already provided the wherewithal for the new station. Without the authority and clout of the military under the looming shadow of World War II, Victoria would not likely ever have realized its dream of acquiring a major airport.

Though the funds may have been available for land acquisition, the process was difficult and complex, with some problems not resolved until long after the base was in full wartime use. Predictably, there were local residents who vigorously opposed the whole project, viewed in some quarters as environmental vandalism—in effect, the destruction of the Saanich Peninsula, "which [to quote one critic] is generally recognized as the most beautiful residential district in Western Canada, and occupied largely by retired people from all over the Empire." Concerns about potentially disruptive noise were expressed by Rest Haven Sanatarium and Hospital and by the Rector of Holy Trinity Church, on the corner of West Saanich and Mills Road. Another critic considered it suicidal folly to put a military airport in close proximity to B.C.'s two major cities, thus almost assuring their swift destruction in the event of a war with Japan.

Every effort was made to counter these objections, and in most cases the negotiations proceeded smoothly. Meanwhile, residents of Sidney began to display a pride in the impressive development next door, especially after the outbreak of war in September 1939. The Crown acquired title to the original airport land package on January 16, 1941; by then, however, it was realized that further extension of the boundaries would be required. Priorities of national defence allowed new construction to proceed without delay, but some of these further negotiations dragged on for another two years. When the Government cheques finally arrived, most Pat Bay farmers seemed to agree that the settlements were fair.

The most troublesome issue of land acquisition erupted on May 14, 1940, when the RCAF base was already in operation; this dispute, documented in eight full pages of the DND journal, was not resolved until January 1942.

North Saanich Consolidated High School, an almost new building (1936), was located at the corner of Centre Road (now Cresswell) and Airport Road, near the southern boundary of the station. In May 1940, the North Saanich School Board expressed concern to Ottawa about the safety of students in the event of a wartime attack on the airport. The trustees' bold proposal that DND should fund a new school was rejected out of hand, if only because it would set a totally unacceptable precedent. The School Board must have grudgingly accepted this Ottawa verdict. However, when it was realized in December 1941—two weeks after Pearl Harbor—that there was an RCAF bomb storage vault dangerously close to the school building, Provincial authorities felt that the school must be closed. An exchange of urgent telegrams revealed that the critical distance was exactly 634 feet—too close for comfort. Ottawa was now forced to assess the cost of taking over the school and funding its replacement.

In an atmosphere of cloak-and-dagger secrecy, a price was soon determined. Luckily for all concerned, DND discovered that the school would provide admirable and badly needed barracks facilities. At the highest federal level, an Order in Council (P.C. 724) transferred ownership of the existing building to the Crown, and allowed North Saanich to receive $43,000 for a ten-acre site and a fine new high school, located some distance to the north of Sidney.

Samuel Brethour became Sidney's original landowner in 1873; his son Julius, who gave the town its name, was an influential backer of the V&S Railway. In the early years of aerial barnstorming, pilots like Harry Brown and Maurice McGregor regularly used the Brethour Farm as the best landing field in North Saanich. Samuel Brethour's grandson, Philip E. Brethour, was among those who had land appropriated for the Patricia Bay RCAF Station. The Brethour name is well remembered at Victoria International Airport: the serene and lovely family graveyard is near the end of Canora Road, and Philip Brethour Park is nearby, on Ocean Avenue.

Most of the essential land for the Patricia Bay Aerodrome and Seaplane base had been assembled in the twelve months following Ottawa's official approval, from March 1938 to March 1939. Early in the spring of 1939, surveying (March 29) and runway grading got under way, and soon the terrain had become even more level than it was before. By summertime, tenders were being called for the surfacing of runways and the erection of hangars. RCAF planners opted to use an asphalt surface on runways; a newspaper item pointed out that the area to be surfaced was equivalent to twenty-six miles of twenty-foot-wide roadway (*Colonist* 1939.08.17: 8). Aprons and high-use areas would be of concrete.

Given the fact that federal Civil Aviation authorities had earlier shown no interest in creating a Vancouver Island airfield, with all progress to date the result of RCAF initiatives, it may seem ironic that the first aircraft to land officially at Patricia Bay was a Department of Transport Lockheed L-12A Electra Junior (CF-CCT), which brought a number of government officials on an inspection tour. The date was August 16, 1939, and things had changed since the airport question had been raised back in 1936. At that date, the Department of Transport did not yet exist, nor had the nation heard much about the Rt. Hon. C.D. Howe, Canada's first and, it seemed, supremely powerful Minister of Transport. By 1939, C.D. Howe was a national celebrity.

Photo by Mrs. Alexander Fraser, courtesy Hugh Fraser

August 16, 1939: DOT Lockheed 12-A Electra Junior (CF-CCT) is the first aircraft to land officially at Patricia Bay, arriving at 10:45 a.m. and departing at 12:45 p.m. A second group would arrive by Lockheed 14 at 3:30 p.m. and leave at 6:00 p.m. The runways have been graded, but not yet surfaced. The fuselage reads "Department of Transport" and the automobile at left reads "B.C. Police." This DOT aircraft, CF-CCT, is now in the Canada Aviation Museum, Ottawa.

There was some chagrin in Victoria that he had chosen not to fly over from Vancouver that August morning, though he would arrive at Esquimalt Harbour by Canadian Airways seaplane two days later. A cynical Victorian might think the Minister was trying to prove that the city did not need land-based TCA service.

Both historic events—the two visiting aircraft of August 16 and Howe's appearance on August 18—were front-page newspaper stories in Victoria. The Minister thought that the Pat Bay runways looked "splendid" as he flew over them en route, and he pointed out that one of them, at 7,000 feet, was now the longest in Canada. [He erred: in Stage I, it was only 5,000 feet.] In addition to its military role, he believed that Patricia Bay would provide an excellent alternative landing field for TCA planes whenever they could not land in Vancouver. But he felt that Victorians had nothing to gain by abandoning seaplanes, which he considered far more economical.

To be fair to the Right Honourable Mr. Howe, Canadian Airways and Canadian Pacific Air Lines would give Victoria admirable passenger service by seaplane over the next four years, and few Victorians thought about Patricia Bay except as a wartime RCAF base. And that airport still seemed a long, long way from the city.

Our next section surveys the early growth of the war-time base, highlighting the pioneer RCAF squadrons and the distinctive aircraft they flew. Of similar interest is the story of the RAF Operational Training Unit, which brought airmen from Britain, Australia, and New Zealand. In keeping with our usual practice, we shall also describe the physical layout of the airport, illustrating some of its more important historical features. Given the fact that Patricia Bay became one of Canada's largest wartime stations, no more than summary treatment is possible in a work of this general nature.

## 111 Coast Artillery Cooperation (CAC) Squadron: October 22, 1939

Neither the surfacing of runways nor the construction of station facilities proceeded as fast as the RCAF might have hoped, but steady progress was made in the months following Canada's declaration of war in September 1939. Meanwhile, an advance detachment arrived from No. 111 (CAC) Squadron, based at Sea Island.

Courtesy George Maude

At 15:11 hours on Sunday, October 22, 1939, F/L George Walter Du Temple and Corporal W.S. (Rusty) Hopper made a pioneer landing in an Avro 626, RCAF Serial No. 225—according to Carl Vincent (1980A: 27) "the first RCAF aircraft and crew to be based on Vancouver Island." Because the runways were not yet completed, they landed on the abandoned CNR railway grade, made serviceable by a roller rented from the B.C. Government.

By Tuesday they were joined by twenty-two men and a truck from No. 111 (CAC), who found accommodation in former civilian homes on the base. On Thursday, October 26, 1939, the RCAF ensign was hoisted for the first time over Patricia Bay Station. A temporary hangar was provided by the Jones barn, seen in the photograph above as only partially covering Avro 626 225. Alas, the small detachment soon found itself without even this one aircraft, as No. 225 was damaged beyond repair in a takeoff accident on November 5 (Vincent 1980A: 29).

In February 1940, 111 (CAC) was advised in Vancouver by Air Commodore A.E. Godfrey that its wartime base would be Patricia Bay, and preparations were made for an early move. On May 15, 1940, the Squadron's four Westland Lysanders and one Grumman Goose flew across the Strait of Georgia, and equipment followed the next day by barge. By May 19, the move was complete (Vincent 1980A: 32). On May 20, 1940, Wing Commander G.A. Mercer assumed command of Patricia Bay Station; the officers' and sergeants' messes were inaugurated the same day.

S/L A.H. (Hal) Wilson

S/L G.W. Du Temple

F/L F.M. McGregor

F.M. McGregor Collection

Unlike Hal Wilson and George Du Temple, Maurice McGregor was not an active member of No. 111 (CAC) Squadron at Patricia Bay; but he was an important Pat Bay pioneer as TCA's senior pilot (see p. 128).

There are many good reasons to shine our spotlight on No. 111 (Coast Artillery Cooperation) Squadron. It is a fine example of an auxiliary RCAF unit that had served B.C. and Canada faithfully through the 1930s; its activities have been very well documented; it was the pioneer resident squadron, having Patricia Bay Station all to itself through July 1940; and it has two important links with civil aviation in Victoria.

Maurice McGregor had risen to the rank of Flight Lieutenant in 111 (CAC) while pursuing his career as bush pilot and airline captain; and when the full complement of 111 (CAC) arrived at Pat Bay in May 1940, it was under the command of Squadron Leader A. Haliburton (Hal) Wilson, who had been Maurice's first flight instructor in December 1928.

## Unscrambling the Patricia Bay Squadrons ▪ 1939 to 1945

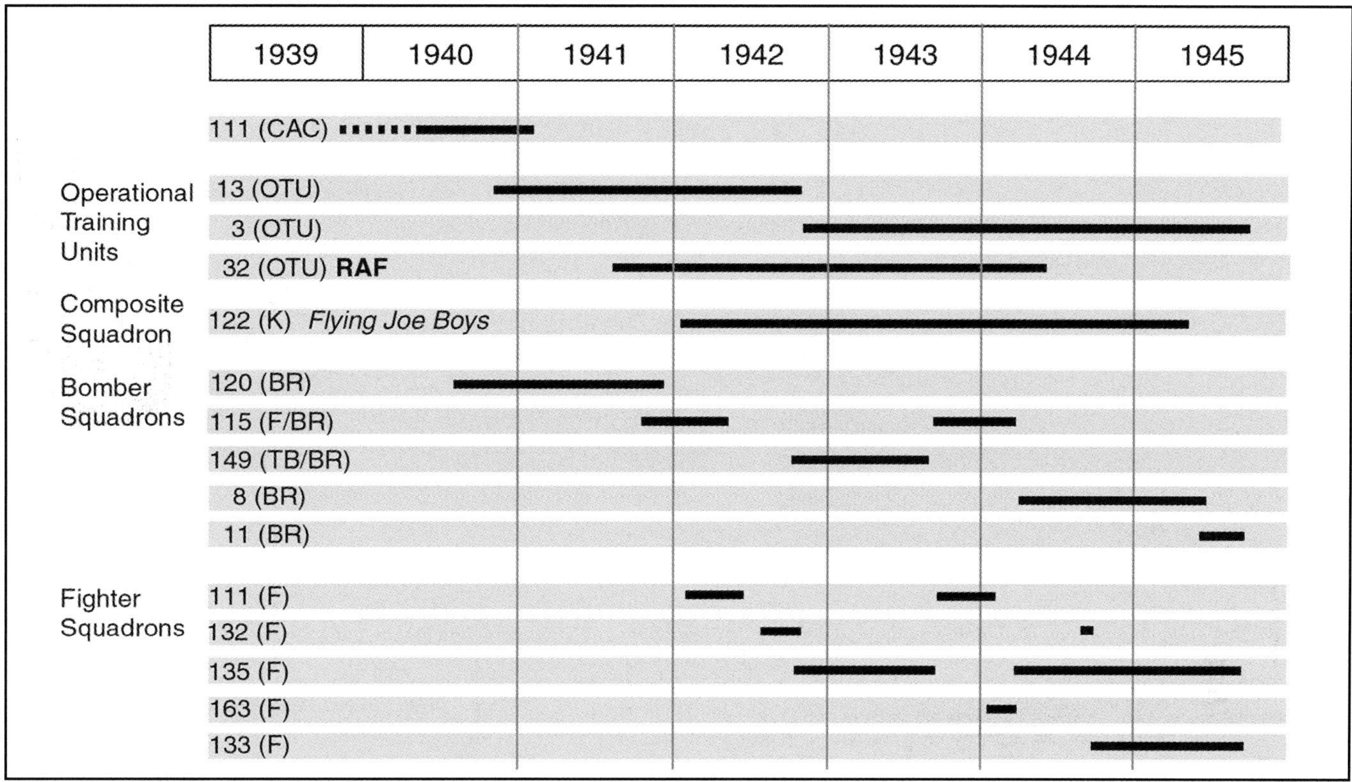


Except perhaps for Pat Bay RCAF veterans, it is a challenging task to sort out the bewildering array of squadrons, training units, and aircraft stationed on the base between 1939 and 1945. The graphic chart above presents a somewhat oversimplified schematic view.

Viewed at an early stage as part of the Commonwealth Air Training Plan, Patricia Bay Station evolved into an unusual hybrid: an advanced training post for both RCAF and RAF, and an operational base with a key defensive role within the Western Air Command. Sometimes the line between trainees and operational flyers gets blurred. Further complicating the picture is the fact that the station was both an airport for landplanes and a major seaplane base. The squadron roster was always changing, as was leadership: Pat Bay had eighteen commanding officers.

Although there was no single "Headquarters Squadron," there were some units that provided continuity. The original 111 (CAC) squadron played that role for over a year, but was then disbanded on February 1, 1941. Many of its personnel and most of its aircraft were absorbed into 13 OTU (Operational Training Unit), which was in turn the forerunner of 3 OTU. Composite Squadron 122 (K), which existed for over three years, was based both on land and sea, performing such a variety of duties that it bore the nickname "Flying Joe Boys." In 1944 Walt Disney dubbed it "The Flying Nightmares," creating for No. 122 Squadron a very distinctive cartoon emblem.

When the RAF detachment arrived, it had its own unit (32 OTU), its own East Camp, and its own aircraft.

If we concentrate on those more continuous elements, it will then be easier to understand the ten RCAF Bomber and Fighter squadrons that arrived and soon left—and, in four cases out of ten, returned later. With the sole exception of 120 BR (Bomber Reconnaissance), whose aircraft included Stranraer flying boats, they all flew landplanes. Not all ten will receive equal attention. Our historical summary excludes the activity of those squadrons at bases other than Patricia Bay—whether on Vancouver Island at Ucluelet, Tofino, Comox, and Port Hardy, or in Alaska at Annette Island and the Aleutians. Their story has been reasonably well documented.

While surveying the Pat Bay squadrons and their aircraft, we provide a brief account of the station's construction, including its historical division into West Camp and East Camp. We also illustrate the important features of the Patricia Bay seaplane base, administered as a separate Marine Section.

Readers who know only the 21st century Victoria International Airport may be pleased to recognize some surviving historic buildings and landmarks that date back to the early years of the Second World War.

Between October 1939 and May 1940, while the small detachment from 111 (CAC) Squadron was temporarily living in the Jones House and other former homes on the base, the construction of RCAF West Camp (above) made steady progress. Located close to the former CNR right-of-way, part of which formed Airport Road (now Willingdon), it eventually included a large steel-and-concrete main hangar, administration building, four wooden hangars, functional living quarters and mess halls, fire hall, workshops, bomb dumps and other facilities. A hospital and officers' quarters were then built on the north side of the airport, close to Mills Road (page 126).

One of West Camp's earliest structures was the permanent double landplane hangar (Building 17), which often appears in photographs alongside a much smaller workshop building (Building 16 on plan). Hangar 17 is still in use over sixty years later as headquarters for 443 Maritime Helicopter Squadron and its Sea Kings. Immediately below is a photograph taken from inside.

During the wartime years at Patricia Bay Station, many photographs were taken against the backdrop of the two buildings enclosed by the broken line on the plan below. Three pictures across the page illustrate that practice. Two show aircraft of the pioneer Squadrons 111 and 120, whereas the third, of 122 (K) Squadron, depicts a later stage in West Camp's development.

DND/PAC AG-311

On August 14, 1940, Air Vice-Marshall G.M. Croil held an aircraft inspection at Patricia Bay. Seen here are a Blackburn Shark, Fairey Battle, and Grumman Goose.

DND/PAC RA-1000028

This Blackburn Shark II No. 501 (and No. 503) arrived at Pat Bay for 111 (CAC) Squadron on July 3, 1940.

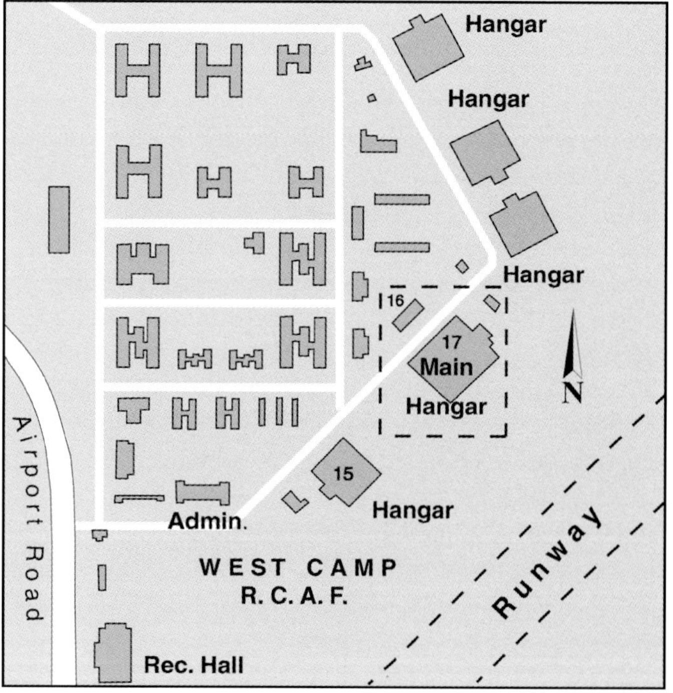

Late 1940: three aircraft stand in front of that conspicuous West Camp main hangar. Left to right, they are a Northrop Delta belonging to 120 (BR) Squadron (see p. 129), and two aircraft from 111 Squadron: Fairey Battle No. 1656, and a Stinson Model 105 (one of two Stinsons recently acquired by 111).

The Westland Lysander, usually built in Hamilton, Ontario, was a familiar aircraft at Patricia Bay RCAF Station throughout the Second World War. Here we see four camouflaged, English-built Lysanders from No. 122 (K) Squadron, whose Pat Bay duty spanned the years 1942 to 1945. In the background we can see two of three wooden hangars in West Camp to the north of Hangar 17; these do not yet appear in the pictures at the top and bottom of this page. All major West Camp construction was complete by mid-1942.

George Maude Collection

Squadron 111 (CAC) arrived at Pat Bay in May 1940 with four Westland Lysanders, and used the "Lizzie" as a versatile support aircraft. Here No. 416 TM-A, is attached to a towing dolly; bomb carriers are visible on its undercarriage. When the report was received of a possible enemy submarine offshore on January 4, 1941, this Lysander 416 stood by for two days equipped with two 250 lb. anti-submarine bombs; but no action took place (Vincent 1980A: 34).

DND/PAC AG-566 (January 11, 1941)

DND/PAC A-100032

Representative aircraft of 111 (CAC) Squadron outside the permanent hangar at West Camp, September 1940. (L–R) Blackburn Shark No. 501, Fairchild 71 No. 647, Blackburn Shark (TM-F) No. 503, and Lysander No. 428.

AG 579 courtesy *Times Colonist*

(Left) RCAF Officers' Quarters near Mills Road, January 1941.

(Below) Construction of Officers' Mess, January 21, 1941.

(Plan below) Area to north of West Camp, near Mills Road: Hospital, Officers' Quarters, Officers' Mess. These standardized wartime wood-frame structures were completed by the spring of 1941.

AG 624 courtesy *Times Colonist*

In July 1940, Victorians became vividly aware of the swiftly growing Patricia Bay Station, and gave it generous financial support. A campaign called "Silver Trail" set out to raise $50,000 in order to buy trainer aircraft. A full-dress RCAF parade was held through city streets. The *Victoria Daily Times* of Saturday, July 20 was filled with this story. A special section entitled "Wings Over Patricia Bay" featured a display of photographs—the different aircraft and the men who flew them.

Wing Commander G.A. Mercer, officer commanding Patricia Bay Airport, explained that this was, at present, purely an operational station. "Most of the training schools are on the prairies and in Ontario because the land there is flat, no danger of running into mountains."

Men taken at "Pat Bay," as they call it, are leading aircraftsmen to be trained in advanced stages of their special field, Wing Commander Mercer explained. Some are air gunners, observers, navigators, wireless operators, fitters, mechanics, and armorers.

—*Times* 1940.07.20: Section 4

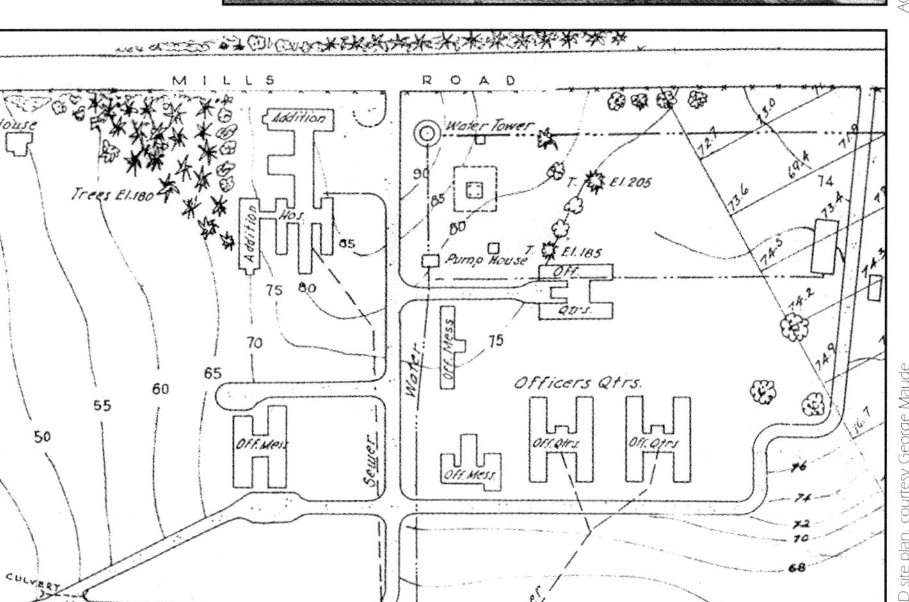


1943 DND site plan, courtesy George Maude

S/L G.W. Du Temple followed S/L A.H. Wilson as C.O. of 111 Sqn. from November 17, 1940 to February 1, 1941.

Two sturdy buildings from West Camp have lasted and have received continuous use into the 21st century. They are the steel-and-concrete landplane hangar (p. 125), built by contractors Carter & Hall, and the three-storey brick Administration Building (shown right in its wartime camouflage), built by Knott & Jones.

DND Photo May 10, 1943: George Maude Collection

*Aerial view of Patricia Bay Airport in 1943 (key appears below).*

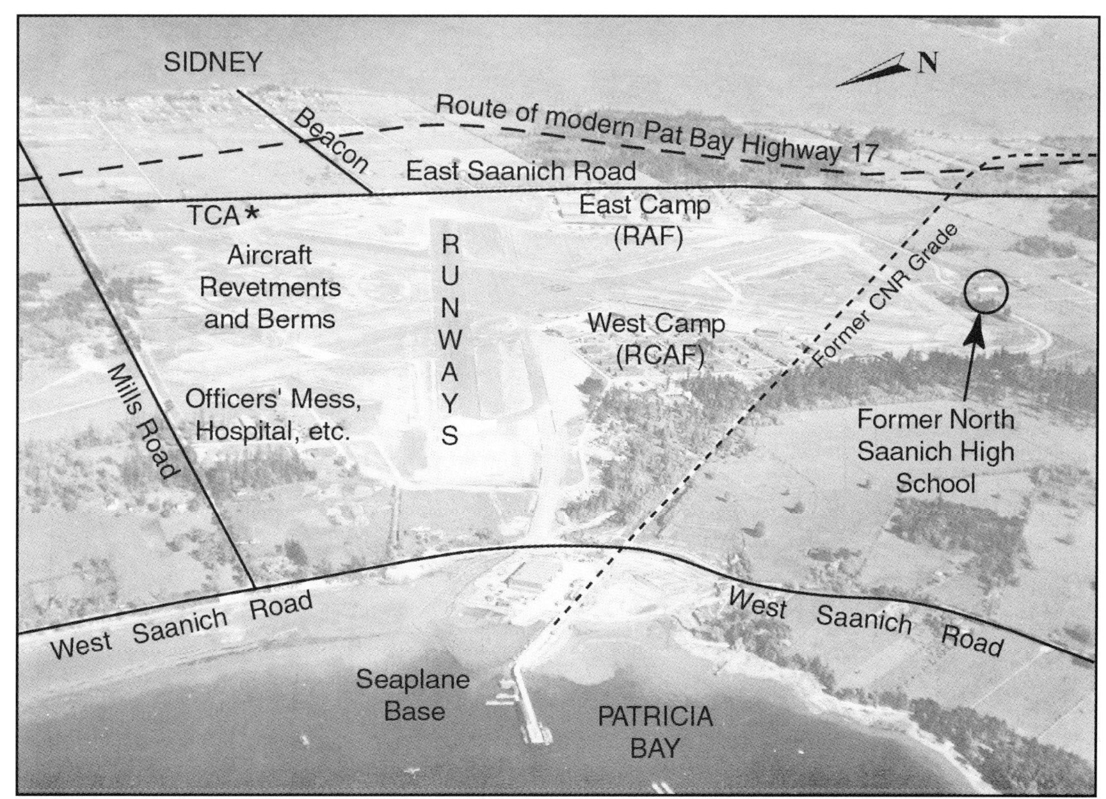

## 1940: A Visitor from Trans-Canada Air Lines & Diverted Flights

Photo by Elwood White

The aircraft above, photographed on the tarmac at Patricia Bay, is a TCA Lockheed 14H, fleet number 29 (CF-TCG). Lockheeds of this type were the first commercial aircraft to land at Pat Bay airport, between June and December 1940.

As TCA's senior pilot, Maurice McGregor had been flying this model on the Vancouver–Winnipeg night run since September 5, 1938. He was Captain of CF-TCF on the first Trans-Canada night airmail run (March 1, 1939). By 1940, he had flown all fifteen Lockheed 14H aircraft in the TCA fleet.

While still on regular TCA service between Vancouver and Winnipeg, Maurice completed four familiarization flights at Patricia Bay between January and September 1940, checking the radio range site and runways (FMM Log 1940.01.08, 06.11, 06.20, 09.05). It was a happy coincidence that he had been a Flight Lieutenant in Pat Bay's RCAF 111 (CAC) Squadron.

On Thursday, June 20, 1940, Captain McGregor twice landed at Patricia Bay Airport in Lockheed 14H (CF-TCE), the TCA prototype that he had ferried to Winnipeg from Burbank, California on May 10–13, 1938. This 1940 pioneer event was noted in the *Victoria Colonist* (see right). On September 5, Maurice made three night landings in another 14H (CF-TCR).

These were not just theoretical exercises: Patricia Bay Airport was now designated as the official alternative field for commercial and other aircraft unable to land at Vancouver Airport on Sea Island. The need for that alternative proved necessary within six months of McGregor's publicized visit.

Between December 14 and December 16, 1940—when Captain McGregor happened to be enjoying several days off duty in Lethbridge—at least five TCA Lethbridge–Vancouver flights were diverted to Patricia Bay, events that received gleeful publicity in Victoria. (Victorians always take smug delight in winter days of brilliant sunshine when Vancouver is enshrouded in fog and gloom.) The first of these landed on Saturday, December 14, 1940, at 11:45 a.m., having left Lethbridge at 8:20 a.m. The plane was flown by Captain George Lothian, and carried five passengers plus flight crew. The second aircraft landed soon afterwards, at 12:30 p.m., with Captain Lindsay Rood, ten passengers, and crew. Both planes were Lockheed 14H models; it is possible that one of them might even have been CF-TCG, shown above.

The *Victoria Times* trumpeted the news on December 14: "TCA Planes from East Make Landings Here." (Well, Lethbridge is quite far east, isn't it? To a Vancouver Islander, any town on the far side of Chilliwack seems more or less beyond Hope.)

These, then, were the very first scheduled commercial aircraft to land at Patricia Bay Airport, even if their schedules had been somewhat compromised. To compensate the dislocated passengers for any inconvenience, TCA agent Wallace Courtney arranged a free brunch in Sidney and passage to Vancouver by CPR steamship. The first commercial flight out of Patricia Bay left for Lethbridge on Monday, December 16, carrying nine passengers who had sailed from Vancouver Sunday midnight on the *Princess Elizabeth*.

It would be two and a half years before a TCA airliner was deliberately scheduled to land at Patricia Bay.

**AIRWAYS PLANE CONDUCTS TEST**

**Initial Landing on Patricia Bay Field Made During Beacon Check**

Piloted by Captain Maurice McGregor, chief pilot of Trans-Canada Airways, a T.-C.A. airliner landed on the airdrome of the Patricia Bay R.C.A.F. station on Thursday, marking the first occasion on which the Defence Department's new landing field has been used by civil aircraft.

With the consent of the Western Air Command, the landing was made in the course of a test flight from Vancouver for the purpose of checking the radio range. The big Lockheed 14H machine, one of the company's trans-continental aircraft, returned to Vancouver the same day.

*Colonist* 1940.06.22: 3

Lockheed Hudsons Mk.I at Pat Bay: October 15, 1941. Four aircraft of this type were flown by 120 (BR) Squadron.

First to arrive at Pat Bay of all its bomber and fighter squadrons, 120 (BR) had formed originally as an auxiliary unit in Regina, on June 1, 1935. At Pat Bay the emphasis was on reconnaissance, since Canada's west coast remained virtually free of hostile intruders.

Flight assignments were still dangerous. Hudson 776 (MX-T), seen in the foreground above, crashed into the sea at Shoal Harbour on February 21, 1942, with loss of aircraft and crew. By then (in December 1941), 120 squadron had left for Coal Harbour, and the Hudsons were being flown by 13 OTU. The bodies of Sgt. Hatfield and P/O B. Hutchinson were recovered by divers, a grim and increasingly common event at the bases of Western Air Command.

The Lockheed Hudson Mk.I was a military export version of the Lockheed 14H, known also as the Super Electra (see photo across page).

Squadron 13 OTU was later active in using various flying boats, such as Consolidated Cansos and Catalinas, but flew only Supermarine Stranraers briefly at Patricia Bay. From 1939 to 1942, the squadron code for 120 (BR) was MX- , as we see from the upper two photographs.

The Northrop Delta Mk.II was one of the most distinctive looking aircraft flown by the RCAF in WWII. Here No. 676 (MX-C) is equipped for land use; this same plane was also used at Patricia Bay on pontoons.

The Supermarine Stranraer was built by Canadian Vickers. No. 120 (BR) Squadron had four at Patricia Bay, but this "Stranny" seen taking off belonged to Pat Bay's Headquarters Squadron, 13 OTU: No. 957 (MK-F).

# 13 Operational Training (OT) Squadron ▪ November 1, 1940

Generally viewed as Patricia Bay's Headquarters Squadron, 13 Operational Training was broadly engaged in flying both landplanes and seaplanes. Here, one of its graduating classes poses in front of Lockheed Hudson Mk.I No. 765, previously on loan to 120 (BR) Squadron (see page 129).

A lineup of Bristol Bolingbrokes Mk.IV from 13 Operational Training Squadron. The nearest aircraft is numbered 9059.

PBG-1302

Bristol Bolingbroke No. 9035 on Main Runway 4 at Patricia Bay Airport.

Formed in Vancouver as a seaplane training school on May 1, 1940, it was renamed No. 13 (OT) Squadron on July 30, 1940, some three months before it arrived at Patricia Bay. It is always called a squadron in definitive RCAF histories, but was widely referred to at Pat Bay as 13 OTU.

In its two-year service at Patricia Bay, from November 1, 1940 to November 1, 1942, this unit flew a wide variety of aircraft, some of which had been inherited from 111 (CAC) and other departing squadrons. In addition to Hudsons and Bolingbrokes, the list included Stranraer and Grumman Goose (see next page), Norduyn Norseman and Northrop Delta.

In November 1942, 13 OTU was disbanded. Most of its aircraft and personnel were absorbed by 3 OTU.

B.C. Aviation Museum

No. 13 (OTU), otherwise known as 13 (OT) Squadron, was a major presence within Patricia Bay's Marine Section. Here are three pictures from the summer of 1942, when 13 OTU aircraft displayed the squadron code MK- (a practice that the RCAF soon abandoned for security reasons).

In the upper two photographs, RCAF Stranraer flying boat No. 916, coded MK-A, is on the concrete apron outside the original seaplane hangar (see next page). In the top photo, as we look eastward towards the airport, we can see the huge travel crane, which operated on an east–west track, and a coffee shop on West Saanich Road. In the middle picture, the same aircraft is in a different position, facing west.

Right: Grumman Goose 942 (MK-H) waits at the bottom of the slipway. The dark building at the top of the ramp is the Marine Office, barely visible in the middle picture.

George Maude Collection

George Maude Collection

Grumman Goose No. 942 (MK-H) at the Patricia Bay slipway.

# Patricia Bay Seaplane Base

Excerpt from DND aerial photograph, 1943

The photo detail above (from page 127) is dated 1943, as is the excerpt from an RCAF station map on the right. Both show one large seaplane hangar in use, and a second (with its own apron) under construction to the south. The wharf and slipways are clearly defined. Airport Road (No. 3), now Willingdon, and the taxiing ramp (No. 4) that led downhill from the main runway should help to orient the images and relate the 1943 complex to the modern waterfront at Patricia Bay.

The original seaplane hangar, long since demolished, used to be visible at the west end of the main runway; see page 138. The second hangar, under construction in 1943, has survived into modern times as the large building that now belongs to the Institute of Ocean Sciences.

A "travel" crane on rails was designed to assist in moving aircraft and other heavy loads. It appears on page 131 and in the photograph lower right.

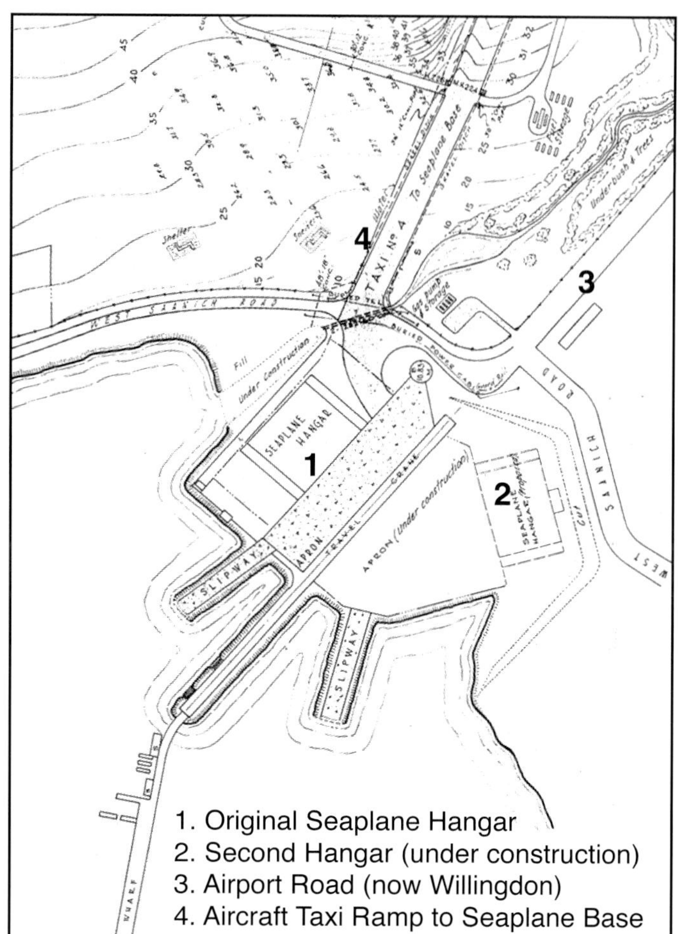

Excerpt from DND Patricia Bay Station map, 1943

B.C. Aviation Museum

B.C. Aviation Museum

B.C. Aviation Museum

Above is RCAF crash boat M-235, named *Huron*. On the right, from the summer of 1942, are two views of crash boat M-232, the *Takuli*. Since the early days of Patricia Bay Marine Section, the importance of these vessels and their crews had been well understood; see page 134. They were state-of-the-art water craft, capable of speeds up to 47 knots.

B.C. Aviation Museum

On the right is a target-tow exercise from April 1942. A seaplane has just been launched from the slipway, as the beaching gear is being wheeled back to the hangar. The aircraft is Blackburn Shark Mk.II, No. 504 (AG-D), from No. 122 (K) Squadron. Four Lysanders from 122 (K) Squadron appeared on page 125. Here we get a glimpse of the large crane that is involved in the exercise. Aircraft refuelling barges can be seen at anchor in the distance.

B.C. Aviation Museum

In this picture, also from the spring of 1942, we can observe some interesting construction activity facilitated by the Marine Section. Huge pipes are being unloaded from a barge for the installation of drainage and sewer systems for Patricia Bay RCAF Station.

B.C. Aviation Museum

Probably the most successful of all Canadian bush-flying aviation designs, the Noorduyn Norseman became very popular in the late 1930s. An 8-seater powered by a Pratt & Whitney 600hp Wasp Jr., it was widely used for general transport purposes in World War II: the RCAF ordered 99, and the U.S. Government bought over 800. At left is a Norseman from 13 (OTU) or 122 (K) Squadron, at Pat Bay in 1942.

George Maude Collection

In this view towards the original Pat Bay seaplane hangar, we see Noorduyn Norseman No. 695 of 122 (K) Squadron; it has come to grief, apparently after a hard landing, and action is being taken to raise it up onto the apron. The incident occurred on October 6, 1942. This particular aircraft, RCAF No. 695, held Norseman serial no. 2, having flown for Mackenzie Air Service of Edmonton as CF-AZA .

## Royal Norwegian Air Force Crash at Pat Bay ▪ March 18, 1941

Within its first year of operation, almost six months before the 1941 arrival of the RAF, Pat Bay experienced a sombre seaplane disaster involving the Royal Norwegian Air Force. Having escaped Norway during the German invasion in 1940, its fledgling student airmen—at first about 120 in number—were now stationed mainly in Toronto, with one small unit temporarily based on Canada's Pacific coast.

Northrop N-3PB

This west coast unit had taken delivery of six new Northrop N-3PB aircraft (#302–307), a custom order for the special needs of the RNAF. The RCAF Northrop Deltas that we have seen already were manufactured by Canadian Vickers, under contract to Northrop; but the 24 new planes for Norway were designed and built by Northrop Aircraft Inc, Hawthorne California (see Hatch 1981, Vincent 1982 and David Maude 1995).

During a brief stay at Jericho Beach Station in Vancouver, there was a tragic accident on February 21, in which Northrop 303 plunged into the sea near Point Atkinson, killing the student pilot, Flight Pupil Erling Jorgensen and his instructor, Harald Kruse, both 25-year-old Norwegians (*Times* 1941.02.22: 16).

For various good reasons, the pro-tem training base was moved to Patricia Bay on March 17, 1940. The very next day, at 10:07 a.m., Northrop N-3PB 305 stalled soon after takeoff from the seaplane base, crashing heavily into Saanich Inlet and catching fire. Pilot Yens Riiser and gunner Kaare Kjos were killed, but Lieut. E. Bjorneby was rescued from the burning wreckage through prompt and courageous action by RCAF Marine Section personnel (*Times* 1941.03.18: 1).

Leading Aircraftsmen Coulton and James of the RCAF Marine Section would receive the MBE for their heroic conduct during this sad event.

(Above) RCAF crash boats on the scene of the first air fatalities in the Patricia Bay region; (below) the salvaged wreckage lies on the ramp at the seaplane base.

## RAF: 32 Operational Training Unit (OTU) ▪ August 9, 1941

### ROYAL AIR FORCE EAST CAMP: MANY OF ITS BUILDINGS STILL SURVIVE

Climatic conditions in this photo, from winter 1942–43, do not show the RAF East Camp at its most appealing.

To the advance party of British airmen who arrived in Victoria on August 9, 1941, the Saanich Peninsula looked like home sweet home. "It's just like England, only with the weather polished up a bit," said Wing Commander B.H. Jones, a Londoner, C.O. of this small detachment.

There was no hush-hush secrecy about the military objective:

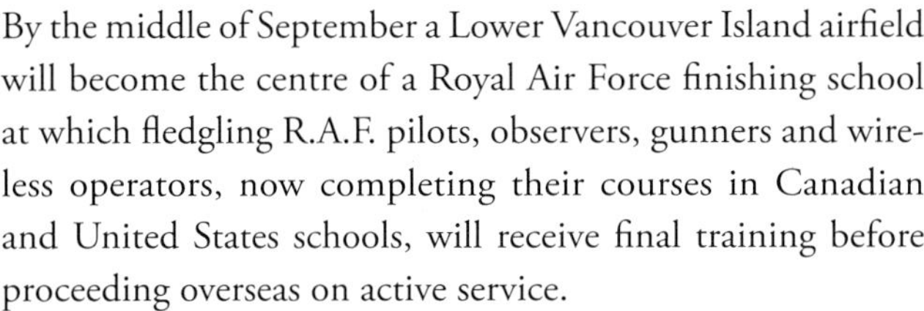

> By the middle of September a Lower Vancouver Island airfield will become the centre of a Royal Air Force finishing school at which fledgling R.A.F. pilots, observers, gunners and wireless operators, now completing their courses in Canadian and United States schools, will receive final training before proceeding overseas on active service.
>
> Spacious hangars, drill sheds and rows of barracks are springing up like mushrooms at the field. Runways are being rushed to completion with heavy pieces of machinery pawing at stubborn obstructions and leveling and hardening the surface of the airport, in preparation for the arrival of the training aircraft.
>
> Known as No. 32 Operational Training Unit, the school, which will be one of the largest in Canada, is not connected with the Commonwealth Air Training Plan.
>
> —*Colonist* 1941.08.14: 1

As if to make the RAF advance party feel even more at home, Victoria had just hosted a Royal Visit from H.R.H. the Duke of Kent, whose ceremonious reception at Patricia Bay is lavishly documented in the Station Diary.

The Royal Air Force would have a lasting impact on Victoria: Jerry Gosley was only one of many British airmen who chose to settle permanently on the Island.

The well-balanced presence of RAF and RCAF squadrons gave Patricia Bay Station a unique profile, and these combined units caused it to become the third largest wartime air base in Canada. Patricia Bay's role was also distinctive: it had a dual responsibility for advanced training and for active vigilance in guarding the Pacific Coast. It was the first line of defence in Canada's Western Air Command, a responsibility that distinguished it from most other training schools across the country.

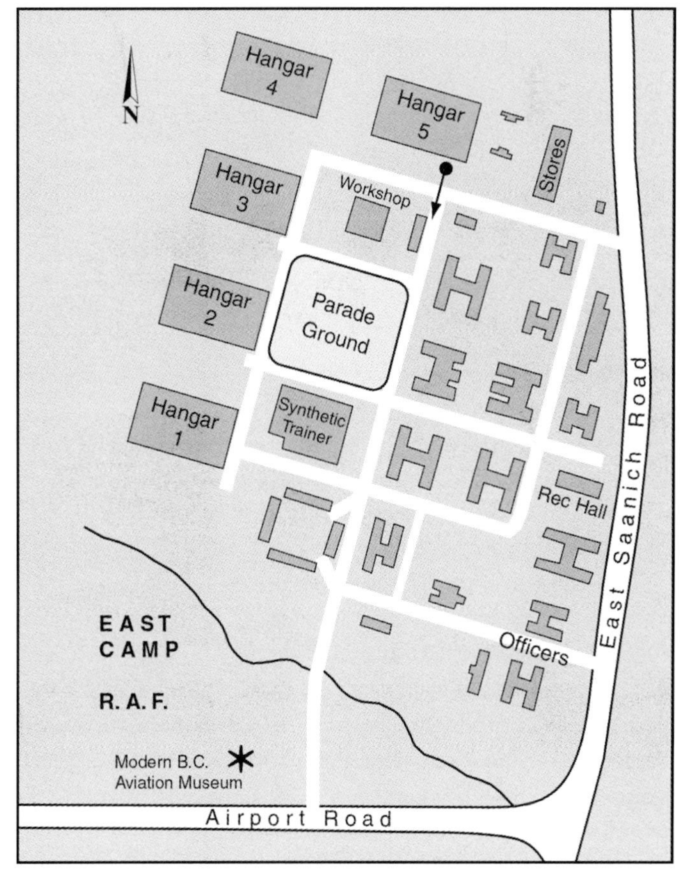


The winter scene above has an opposite orientation to the East Camp plan below. The arrow on the plan marks the spot where the photographer was standing, at the corner of Canso Drive and Harvard Road, looking over the parade ground toward Mt. Newton.

It is a gracious historical tribute that the roads of the modern airport all carry the names of World War II aircraft, with a blending of RCAF and RAF traditions.

PBG 1437

Bristol Beauforts of RAF 32 Operational Training Unit at Patricia Bay: January 20, 1942.

Not long after the first major detachment of RAF servicemen arrived at Pat Bay in September 1941, heavy rains turned East Camp into a muddy quagmire, a situation that prevailed throughout a rather dismal winter. In order to boost morale, two creative young British airmen produced a fine monthly newspaper, *The Patrician*, whose inaugural issue appeared in October. Their names were Jerry Gosley and R.D. Hilton Smith, who were destined to be future pillars of Victoria's cultural life, as "Smile Show" producer and Adelphi Bookstore proprietor. Before long, Gosley—a ground-crew trainee—would be spending most of his time and energy presenting zany variety shows in the East Camp recreation hall. The fame of these productions soon spread far beyond Pat Bay, to Seattle and even to Hollywood.

For an entertaining biography of "Smile Show" wizard Jerry Gosley, with pictures of his days at Pat Bay in RAF 32 OTU, see John Windsor, *Nowhere Else to Go* (Sidney: Gray's Publishing Ltd., 1964).

In that same month of October 1941, shipments of aircraft for RAF 32 OTU began to arrive regularly. The first lots were two-engine Bristol Beauforts manufactured in St. Hubert, Quebec, like those drawn up on the tarmac above. The Beaufort was a reliable and modern general reconnaissance torpedo bomber. The picture at the top of the page was taken in January 1942; the two aircraft nearest the camera, N 1006 and L 9967, had been delivered to Patricia Bay on October 16 and October 29, 1941 respectively. (Useful facts such as this are duly recorded in the diary of 32 OTU's C.O., which is extant.)

Less than a month after that photograph was taken, on February 16, 1942, Beaufort N 1006 (the nearest above) crashed into the sea eight miles southeast of Pat Bay, beyond James Island. P/O (later S/L) J.A. Piddington and his crew were lucky to escape injury; their Beaufort sank, but was later salvaged. The RAF unit had already suffered one death from pneumonia and another from acute appendicitis. The months ahead would witness an appalling loss of life from training accidents. To prepare young pilots for active service in a very short time is always fraught with danger, but RAF 32 Operational Training Unit was particularly unfortunate in this regard.

By July 31, 1942, Bristol Beauforts were being swiftly replaced by Handley Page Hampdens. Aircraft strength at that time was recorded as 12 Beauforts, 22 Hampdens, 9 Avro Ansons, and 8 Airspeed Oxford trainers, for a total of 51. By 1944 total aircraft strength would reach 119, including several Beechcraft and Dakotas.

Staff strength for 32 OTU in July 1942 totalled 1,197 (officers and other ranks), including 33 from the RCAF. In a separate category are listed 119 "Students" (38 officers and 81 NCOs). Of these, 36 were RAF and 43 RCAF, with 29 from the Royal Australian Air Force and 11 from the Royal New Zealand Air Force.

This Beaufort was photographed on April 26, 1942. On June 20 of that year, Beaufort 6484 was damaged on takeoff during a fruitless attempt to pursue the Japanese submarine that had shelled Estevan Point.

For the most part, aviation historians give the Handley Page Hampden only a lukewarm review: it was fast (254 mph) and manoeuvrable, but was so narrow and cramped that its four-man crew were virtually immobilized. For this reason it was dubbed the "Flying Suitcase"; at Patricia Bay, it is more bluntly remembered as the "Flying Coffin." In particular, it seemed to have a tendency to yaw and lose stability while banking. Perhaps this idiosyncrasy posed a challenge to student pilots that could be more easily overcome by veterans.

Whatever the reason, accidents in Hampden aircraft account for the majority of Pat Bay's fatalities in World War II. There are 37 such crashes on record, often involving a four-man crew. During the war, over 100 air crew members lost their lives. Royal Oak Burial Park contains the graves of 58 Commonwealth airmen, from Great Britain, Canada, New Zealand, and Australia. This is a poignant reminder of the fact that not all wartime sacrifices occur on the field of battle.

At the same time, we must keep in mind that about ten thousand trainees passed through Patricia Bay Station during the years 1940 to 1945; a great many of these were young men from 32 Operational Training Unit.

It is hard to give precise figures for wartime Patricia Bay. It was the third largest air base in Canada, and was apparently able to train about 3,500 at any one time. The total complement may have peaked around 5,000.

RAF 32 OTU was reassigned to Comox in May 1944, and all its personnel soon left. Even after that major move, the Patricia Bay Station Diary lists the base's strength, on September 30, 1944, at 3,139. One should note that this number included 321 RCAF airwomen and 8 Nursing Sisters, as well as 112 civilians.

Handley Page Hampden P 5428 from RAF 32 OTU, Patricia Bay. Of 96 Hampdens delivered to 32 OTU, only 49 were on hand in January 1944.

May 16, 1943. RAF Hampden (AJ 994) from 32 OTU drops a torpedo into Saanich Inlet, off Pat Bay. This aircraft crashed and burned on Galiano Island, September 28, 1943, killing Sergeant Herbert D.W. McLeod of the Royal New Zealand Air Force, who was flying solo.

A Hampden of 32 OTU burns just off a Pat Bay runway. There were two fatal crashes of this type in 1943: May 23 (AN 142) and September 17 (AN 146).

## 115 (F/BR) Squadron: Bolingbroke and Ventura ▪ 1941–42, 1943–44

Squadron (F/BR) 115 was among several that were stationed more than once at Pat Bay. Between October 15, 1941 and April 25, 1942 it was designated a fighter squadron, though it flew only Bristol Bolingbrokes. In August 1943 it returned as 115 (BR) from Annette Island, Alaska, swapping its Bolingbrokes for Lockheed-Vega Ventura G.R. Mk.Vs. It flew these from Patricia Bay on anti-submarine duty until March 16, 1944, before a final five-month posting in Tofino.

Upper left: a row of six 115 (BR) Bolingbroke Mk.IV aircraft is lined up beside main runway 4 (now 09-27); nearest the camera is 9140-P. (Note the large seaplane hangar.) The photo is dated August 19, 1943. If that date is right, the transfer to Venturas must have been in progress.

Lower left: this is 115 (F) Sqn Bristol Bolingbroke Mk.I (710), during the Squadron's first posting to Patricia Bay. That type was flown only until December 1941.

PBG-4150

The picture at right is dated August 18, 1943, exactly one day before the large photo at the top of the page. It apparently shows the arrival of Lockheed Venturas 2227 and 2228 for the use of 115 (BR) Squadron. There are Walt Disney insignia near the rear doors of both aircraft. We look north toward the control tower.

PBG-4251

Lower right: on September 17, 1943, less than a month after delivery, this 115 (BR) Sqn Ventura (No. 2223) received considerable damage in an undignified landing. Behind the RCAF roundel is a small figure of Mickey Mouse; all these 1943 Lockheed Venturas displayed Disney characters. We look northwest, past West Camp, to the Officers' Quarters and water tower near Mills Road.

## 149 (TB/BR) Squadron: Beaufort and Ventura ▪ 1942 to 1943

PBG-1483

Shown on this page are the last three bomber squadrons in order of initial Patricia Bay starting dates (see chart, p. 123): 149 (TB/BR)—October 26, 1942 to August 16, 1943; 8 (BR)—March 20, 1944 to May 25, 1945; and 11 (BR)—May 20 to September 15, 1945.

RCAF "Sea Wolf" Squadron 149 was formed at Pat Bay in October 1942, just over a year after the arrival of RAF 32 OTU. Originally it was a Torpedo Bomber unit, meant to counter the Japanese naval threat from the Aleutian Islands. Until July 1943 its pilots flew Bristol Beaufort Mk.I bombers of the type seen above—the only RCAF home unit to do so. These were identical to Pat Bay's RAF Beauforts, many of which were transferred to 149 (TB) by 32 OTU. In July 1943, 149 (TB) was re-designated Bomber Reconnaissance and issued Lockheed-Vega Venturas in preparation for anti-submarine duty at Annette Island, Alaska and Terrace, B.C. [Source: Kostenuk and Griffin (1977: 66); Douglas (1986 II.661) dates 149 (TB) Squadron's formation to October 1, 1942.]

## 8 (BR) Squadron: Lockheed Ventura Mk.V ▪ 1944 to 1945

From March 1944 to May 1945, 8 (BR) Squadron, like 115 (p. 138) and 149 (top of page), flew Lockheed-Vega Ventura G.R. Mk.Vs, as seen here. This aircraft was a military version of the Lockheed 18 Lodestar flown by TCA.

## 11 (BR) Squadron: Consolidated Liberator ▪ May–Sept., 1945

After VE-Day, this distinguished Atlantic Coast RCAF squadron was transferred to Western Air Command; its Consolidated B-24 Liberator G.R. Mk.VI bombers (above) served at Pat Bay from May 31 to September 15, 1945.

Patricia Bay RCAF Station, 1943. East Camp is at the top; West Camp is centre right. The main east–west runway extends from the upper left of the photograph to the lower right.

In the earliest days of Pat Bay Airport, the three runways were identified as 2, 3, and 4; No. 4 was the main east–west runway. Soon they were given conventional designations based on the earth's magnetic pole, so as to match aircraft compass bearings.

Runway terminology can seem mysterious, because the final digit is removed and the result rounded off. Thus Patricia Bay's main east–west runway became 08-26, in the 1980s revised to 09-27 after a minor shift in the magnetic pole. (Compass bearings of 090-270 would be exactly east–west, in magnetic terms.) The three runways shown above, reading clockwise from the lower right, are now designated as 09-27, 13-31, and 02-20.

PB.G. 3250

View southwest toward West Camp, from the vicinity of the control tower (November 5, 1941).

PB.G. 1312

View east toward the hangars of West Camp, from the vicinity of Airport Road (December 16, 1941).

## 3 Operational Training Unit (OTU) ▪ 1942 to 1945

Successor to 13 Operational Training Unit as Patricia Bay's RCAF Headquarters Squadron, 3 OTU was particulary active on the waters around southern Vancouver Island, boasting an enviable variety of the closely related Consolidated Cansos and Catalinas.

The Catalina was exclusively a flying boat, while the Canso was equipped with an undercarriage for either land or water use. This very popular aircraft was manufactured by Boeing Aircraft of Canada, in Vancouver, and by Canadian Vickers, in Montreal.

The three pictures on this page all show 3 OTU Cansos. At top is a classic image of three Cansos taking off as they leave Cowichan Bay, with Mt. Tzuhalem in the background.

The lower photographs show Canso 11025 at the Pat Bay Seaplane base, and Canso No. 11083.

Photo by George Maude

Photo by George Maude

## 135 Fighter (F) Squadron: Hawker Hurricanes ▪ 1942 to 1944

135 (F) Squadron Hurricanes on the taxi ramp of Patricia Bay's main runway; at upper right, are the station hospital and officers' quarters near Mills Road.

Though not the first fighter squadron at Pat Bay—that honour belongs to 111 (F)—135 (F) served by far the longest of any fighter squadron on the station.

It had been formed at Mossbank, Saskatchewan on June 15, 1942. During its first Pat Bay posting, from October 8, 1942 to August 14, 1943, it flew only Hawker Hurricanes. After a stint with US Alaskan Command at Annette Island, it returned via Terrace to Pat Bay with Hurricanes on March 12, 1944; but then it flew Curtiss P-40 Kittyhawks from May 1944 to September 10, 1945.

Our plan is to view all Pat Bay's Kittyhawk squadrons in one group.

Here we see only the Hurricanes of 135 Squadron; all were Mk.XII.

This Squadron's first tour of duty at Patricia Bay coincided with the appointment of the station's twelfth commanding officer, Group Captain Gordon Roy McGregor, DFC, (below) himself a Hurricane fighter-pilot hero of the Battle of Britain. G/C McGregor was Patricia Bay C.O. from April 1, 1943 to February 5, 1944. After the war he would become President of Trans-Canada Air Lines and Air Canada from 1948 to 1968.

Three of 135 (F) Squadron's Hurricanes in flight. The top middle aircraft, 5405, was damaged in a landing accident on February 2, 1943 (see next page).

*Times of London, ca. 1940*

Right: Montreal native Gordon Roy McGregor, CC, OBE, DFC (1901–1971), is a member of Canada's Aviation Hall of Fame. Above is Group Captain McGregor's personal Hawker Hurricane, likely the only one with a spinner and no guns.

Many RCAF squadrons in World War II adopted tough-sounding nicknames. Before it left Saskatchewan, in July 1942, 135 (F) had dubbed itself the "Bulldog" Squadron; therefore the cartoon image of a menacing canine pugilist often appears near the nose of a Squadron 135 aircraft.

Squadron Leader W.C. Connell was C.O. of 135 (F) from July 19, 1942 to October 30, 1943—and thus for all of his Squadron's first posting at Pat Bay. Here he poses alongside a Hawker Hurricane that sports the familiar cartoon emblem. Better still, S/L Connell holds on a leash the Squadron's actual bulldog mascot "King," who looks slightly nervous as he sits aboard the Hurricane's port wing. Behind the dog, we see that the aircraft is named "Thelma," but we have no record of her serial number.

PBG 5603

PBG 5601

At first glance, we may not notice that this bucolic landscape on Saltspring Island is the scene of an aviation accident. It was on November 30, 1942, that Hawker Hurricane 5425 W of 135 (F) Squadron made a forced landing on that field.

While flying over Saltspring—the largest and closest of the major Gulf Islands, situated to the north of Patricia Bay—Flight Sergeant Wheeler encountered complete engine failure, but was fortunate enough to make a commendable emergency landing. Hurricane 5425 was salvaged and rebuilt.

Earlier that same autumn, late in the evening of October 4, 1942, there had been a fatal accident on Saltspring Island: RAF Hampden AN 105 crashed into Mt. Tuam, killing the lone occupant, student pilot P/O Norris Thomas, who had been performing training exercises.

PBG 3850

Several months later, Sergeant Binion's second solo flight in Hurricane 5405 (seen aloft on page 142) had this unfortunate and embarrassing result. Within hailing distance of the Pat Bay Control Tower, the aircraft completed a spectacular curving skid, followed by an acrobatic bottoms-up salute for the traffic controller.

The date was February 2, 1943.

## Curtiss P-40 Kittyhawk Squadrons: 111, 132, 163, 135, 133

PBG 1436

**Curtiss P-40 Kittyhawk Years at Pat Bay**

| | |
|---|---|
| **111** | January to June 1942 |
| | August 1943 to January 1944 |
| **132** | July to October 1942; August 1944 |
| **163** | January to March 1944 |
| **135** | May 1944 to September 1945 |
| **133** | August 1944 to July 1945 |

Pioneer traditions of 111 (CAC) Squadron lived on in 111 (F), but this was an entirely new unit, formed at Rockcliffe, Ontario on November 3, 1941. (The original 111 had been re-designated as a fighter unit in the summer of 1940, but never received fighter aircraft.)

The top photograph shows Curtiss P-40 Kittyhawks (Mk.I) of 111 (F) refuelling at Pat Bay on January 20, 1942—two days after their arrival. Nearest to the camera are AK 905 and AL 190. Alone among RCAF squadrons, 111 was required to share muddy East Camp with the RAF. On March 17, 1942, the unit was adopted by the West Saanich First Nation, who presented 111 (F) with a Thunderbird totem and gave the Squadron its proud *Thunderbird* nickname (Vincent 1980B: 70–71).

From June 3, 1942 until early August 1943, 111 (F) Squadron served under US Alaskan Command, most notably at Umnak Island in the Aleutians.

Stationed again at Patricia Bay until January 15, 1944, 111 (F) acquired additional Kittyhawks, both Mk.IV and Mk.I. These aircraft were left behind at Pat Bay when the Squadron headed overseas on January 20, 1944.

For a clear, authoritative, and well-illustrated history of this 111 (F) squadron, see Vincent 1980B.

After returning to Patricia Bay with Hurricanes in March 1944 (see page 142), 135 (F) Squadron flew Curtiss P-40 Kittyhawks from May 1944 until September 1945.

Above we see well-known Victorian Dave Duncan of 135 Squadron on the wing of his P-40. In the photo to the right, other Mk.IV aircraft of 135 Squadron are drawn up at Patricia Bay. The bulldog emblem is visible on the nearest Kittyhawk, which is probably 858 J.

## George Maude's Second World War P-40 Kittyhawk, No. 1034

Long-time resident of Saltspring Island and Patricia Bay, George Maude served on RCAF crash boats in World War II, and soon afterwards began a long career at Pat Bay Airport with TCA and Air Canada. He is perhaps the foremost authority on the history of the wartime base and its subsequent role in civil aviation. When many wartime aircraft were declared surplus in August 1946, George had the foresight to buy Kittyhawk 1034 for $50, and two Bolingbrokes (9892 and 9104) for $50 and $35.

Above is P-40E, Mk.IA Kittyhawk No. 1034 in its heyday. Received by the RCAF on October 16, 1941, it saw action from Halifax to the Aleutian Islands before its final posting with 133 Squadron at Patricia Bay in 1944–45.

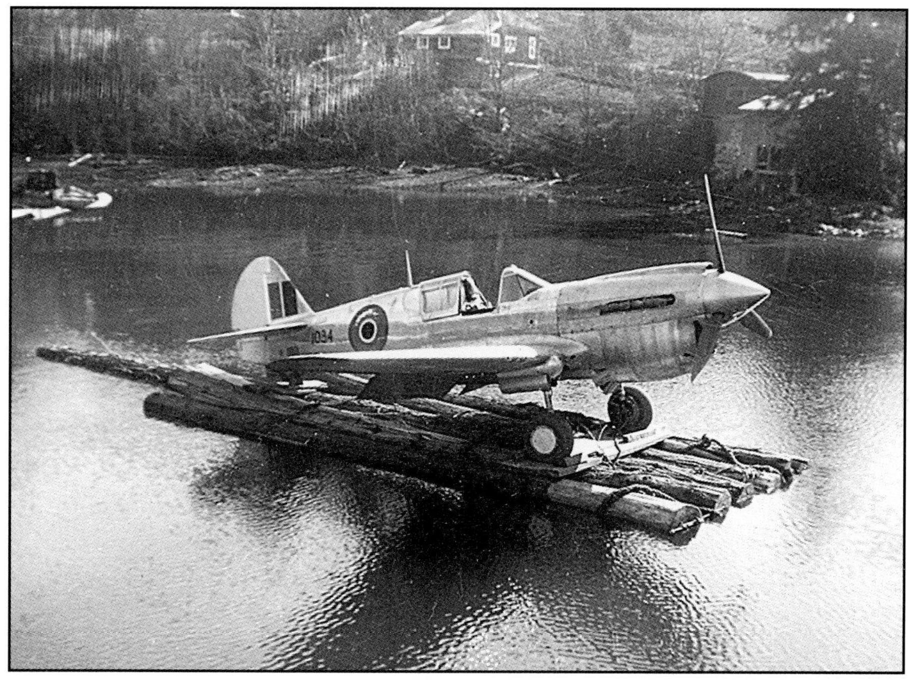

Together with the Bolingbrokes, the P-40 was taken in 1946 by log raft to Fulford Harbour on Saltspring Island, as seen in the two pictures, left. George Maude's father was Captain of the M.V. *Cy Peck*; their family home was very close to the ferry terminal seen in the bottom photograph. There the Kittyhawk was carefully kept in storage until returned in July 1974, via the BC Ferry *Saltspring Queen,* to Victoria International Airport. Restoration was carried out in Hangar 15, in space provided by Bill Sylvester's Victoria Flying Services.

Because No. 1034 consists almost entirely of factory stock, it is credited as the most original P-40 in existence today, with a value almost beyond price. Except for the former machine guns and a new pair of vintage wings, the aircraft is virtually unchanged from the day it left the RCAF.

In 1964, George Maude donated Bristol Bolingbroke No. 9892 to the RCAF, having sold No. 9104 years earlier to a Saltspring friend. Bolingbroke 9892 is now on permanent exhibit in the Canada Aviation Museum in Ottawa. A photograph of this aircraft is included, with a tribute to George Maude, in our final chapter.

## De Havilland Mosquito F.B. Mk.26: 133 (F) Squadron "Falcon"

Photo by Neil A. Macdougall

In the spring of 1945, a group of 133 Squadron's Mosquito fighter-bombers is clustered not far from Patricia Bay's 21st century passenger terminal.

Between March and April 1945, 133 (F) Squadron was converted from P-40 Kittyhawks to de Havilland Mosquito fighter-bombers. Canada's west coast was then under attack by Japanese unmanned incendiary balloons; the Mosquitos, with their superiority in speed and altitude, were expected to have a better chance of interception.

This did not happen. Early in 1945, two fire balloons had been shot down near Pat Bay by P-40s from 135 (F) Squadron, one of which (858) may be pictured at the bottom of page 144. In contrast, the Mosquitos of 133 Sqn scored no victories. The Squadron was disbanded on September 10, 1945; the aircraft were shipped to China.

133 (F) Squadron's Mosquito fighter-bombers, Mk.26. Nearest to the camera is KA112 L.

No doubt, fighter-pilots have always known how to relax when off duty. However, life must have seemed sweeter than usual on a warm day in the summer of 1945, when Germany had been defeated and the war against Japan was winding down. These men at Pat Bay are from 133 and 135 Squadrons, and their aircraft are P-40 Kittyhawks and de Havilland Mosquitos, respectively.

The scene is the Patricia Bay Seaplane Base, near the end of World War II. The original seaplane hangar and Marine Office are in the background. Mosquito fighter-bomber KA126 C, 133 (F) Squadron, is being loaded onto a scow in order to become the main attraction at a Vancouver war-bond rally. (The scow's wooden uprights have been hastily reduced by means of a crosscut saw.)

Student pilots were often able to walk away from horrific-looking crashes. On April 9, 1942, Bristol Bolingbroke 9036 (AN-L) of 13 OTU was damaged beyond repair, after bouncing off the runway on a routine landing.

George Maude Collection

Patricia Bay, 1944. When the C.O. first caught sight of this neatly symmetrical but highly unorthodox landing formation, involving 135 Squadron's Curtiss P-40s 864 F and 858 J, he is said to have exclaimed sarcastically, "What do they think that place is, a bloody dartboard?"

For all their virtues, the celebrated Kittyhawks had a strong tendency to end up in a nose-down position if the brakes were applied too hard.

DND PBG 5724

On June 1, 1945, the last RCAF crash at Patricia Bay involved this de Havilland Mosquito (KA 111) from 133 (F) Squadron, flown by Harold S. Lisson. The aircraft lost one of its engines while landing, and ended up in a field near Mills Road. The house in the background is the former Matson home, later demolished, which was near the Officer's Mess.

If there is one thing that an aviation enthusiast hates worse than a damaged aircraft, it is an aircraft in mint condition that is slated for the scrap heap. That was the fate that lay ahead for these almost new Vancouver-built Catalina flying boats, seen at Pat Bay in 1945 before their demolition by the Victoria firm of Capital Iron. A few visionaries saved the occasional war surplus airplane, but the great majority came to a sadly wasteful end.

## Hollywood Invades Patricia Bay—Twice

Hollywood came to Patricia Bay in 1942, with the filming of *The Commandos Strike at Dawn*, a super-patriotic boost to the war effort that starred Paul Muni, Lillian Gish, Sir Cedric Hardwicke, and Robert Coote. Saanich Inlet became a quite convincing Norwegian fiord, and Pat Bay was the Nazi air base that had to be captured by means of a stealthy British commando raid. All Victoria basked in the glitter and excitement.

The ever-resourceful and dauntless Jerry Gosley invited the leading actors to perform at a special "Smile Show" in East Camp recreation hall. Paul Muni demurred, but RAF 32 OTU's stage was graced by the presence of Sir Cedric Hardwicke, Lillian Gish, and Robert Coote (who created the role of Colonel Pickering in Lerner and Loewe's *My Fair Lady*). No doubt Jerry was able to hold his own with that stellar trio.

B.C. Aviation Museum

B.C. Aviation Museum

This was not the usual appearance of Patricia Bay Station. As one might suppose, ersatz Nazi military insignia were highly coveted souvenir items.

B.C. Aviation Museum

Patricia Bay's second taste of Hollywood was the 1945 MGM film *Son of Lassie*, which starred Peter Lawford, June Lockhart, Nigel Bruce, Donald Crisp—and Lassie. At the conclusion of filming, on August 29, 1944, Director S. Sylan Simon (centre) presented the station with a new mascot—Lassie's daughter, Miss Patricia. At left is Air Vice-Marshal F.V. Heakes with Lassie; on the right is Patricia Bay's Commanding Officer, Group Captain Basil D. Hobbs.

## PART SIX

# YEARS OF TRANSITION AND DEVELOPMENT

## 1943–1958

B.C. Archives I-26684

# 15. TCA Begins Regular Passenger Service to Victoria ▪ 1943

B.C. Archives I-26678

Eagerly anticipated since early spring, the moment arrived at 11:25 p.m. on Sunday, June 6, 1943: Victoria was now officially the western terminus of Trans-Canada Air Lines' trans-continental service. Proudly stepping down from the TCA Lockheed Lodestar was Victoria Member of Parliament Robert W. Mayhew, a dedicated aviation booster, who had flown aboard the aircraft all the way from Ottawa, after enjoying breakfast in the Chateau Laurier at 8 a.m. that morning. From Lethbridge to Victoria the flight crew was Captain Don Brady, First Officer Norman Ramsay, and Stewardess Mina Wood.

Although Victorians were overjoyed to be connected at last to Canada's national airline, after six years of frustration, there was still a major shortcoming in a galling proviso now imposed: would-be local passengers between Victoria and Vancouver were forbidden to fly only that short leg (in either direction). As a result, the twice-daily TCA Lodestars were often almost empty on these segments, though the short-haul demand was always high.

The Hon. C.D. Howe was a staunch supporter of this controversial regulation. The policy was established by Ottawa's Board of Transport Commissioners as a concession to Canadian Pacific Air Lines, which had strongly opposed TCA's extension to Victoria. Since 1942, CPAL had enjoyed exclusive rights to the Victoria–Vancouver route on a licence inherited from its predecessor, Canadian Airways, duly granted that privilege on March 1, 1939.

B.C. Archives I-26679

On this page we see two views of Lockheed 18 (CF-TCT), fn 42, in front of the original TCA Victoria Terminal, just off East Saanich Road, between Beacon Avenue and Mills Road. Long gone from this location, the tiny wooden structure survives as a well-disguised residence in Sidney.

On the previous page, where TCA employees are loading tulips onto the same aircraft, there is another good view of the original terminal. TCA transcontinental air freight service did not become big business until March 1, 1948 (*Colonist* 1948.07.28: 21).

# 16. Canadian Pacific Arrives . . . and Soon Departs ▪ 1945–1946

**December 4, 1943:** a ceremony at Canadian Pacific Air Lines' View Royal terminal welcomed the first Barkley-Grow 8-passenger seaplane, which replaced the DH Rapides that had served Victoria since April 1939 (p. 115). CPAL owned six Barkley-Grows, which flew from Esquimalt Harbour until July 31, 1945. By that time, the frequency of daily round trips had increased from five to nine.

**August 1, 1945:** Canadian Pacific began Vancouver–Victoria passenger service on wheels from Patricia Bay with Lockheed 14s. Bill Taylor took this photograph of CF-CPK at the Pat Bay Terminal on July 12, 1946. The number of daily round trips was now reduced from nine to seven, but the 14-passenger Lockheeds increased the daily capacity from 144 to 196. Victorians were really flying.

W.W. Taylor Collection

Photo by Neil A. Macdougall

**July 27, 1946:** the popular Douglas DC-3 became the second type of aircraft flown by Canadian Pacific Air Lines in its brief fifteen months at Patricia Bay. DC-3 flights would continue until October 31, 1946, when Canadian Pacific was required to discontinue its profitable Victoria–Vancouver service.

July 28, 1945: Canadian Pacific Air Lines announced that its Barkley-Grow T8P seaplanes operating between Vancouver and Esquimalt Harbour would be replaced on August 1 by a land-based service between Vancouver and Patricia Bay Airport (*Colonist* 1945.07.29: 14). Canadian Pacific would now share with TCA the small passenger terminal off East Saanich Road, a tiny wooden building barely visible in the photos above.

A year later, CPAL replaced its Lockheed 14s with 28-passenger Douglas DC-3s, one of which (CF-CPX) is in the lower picture, at the Pat Bay terminal in August. This was a short-lived upgrading, destined to last for barely three months. In 1946 TCA was granted an unrestricted licence for the Victoria–Vancouver run, and Canadian Pacific was no longer allowed to compete. The last CPAL flight left Patricia Bay at 8:15 on October 31, 1946; its Victoria ticket office was closed, and some staff members transferred to TCA (*Colonist* 1946.10.11: 3). It was the end of an era for aviation in Victoria.

James Richardson's Canadian Airways and Grant McConachie's Canadian Pacific Air Lines had been flying the Victoria–Vancouver route since 1931, providing admirable service. Public pressure forced Ottawa to rescind the restriction on TCA, a popular move. In this case Victorians did not side with the loyal underdog, although they would welcome the return of CP Air to Victoria International Airport three decades later.

This fifteen-month service in 1945–46 was not the only time that CPAL or its predecessor had flown on wheels between Victoria and Vancouver. In a letter to Elwood White dated March 9, 1971, Gordon Ballentine reported that he had regularly flown a land-based aircraft from Pat Bay to Vancouver in 1941 and 1942: "In the fall of 1941 Canadian Airways Limited was required to provide a Victoria–Vancouver air mail connection with TCA's 6 a.m. departure from Vancouver to the east. There was no way a seaplane could do this job [at night] so a DH Dragonfly, CF-BPD, was used on wheels. Vancouver Airport to Pat Bay Airport [and return]. We were only permitted to carry mail, no passengers . . . ."

# 17. Trans-Canada Air Lines at Patricia Bay Airport ▪ 1946–1958

Photo by Elwood White

November 1, 1946. Arrival at Patricia Bay of the first TCA Douglas DC-3.

Note that TCA's DC-3 service was inaugurated with that company's first short-haul passenger flight to Victoria, the very day after Canadian Pacific left the scene. This DC-3 above, CF-TDK, may not have been the actual pioneer on November 1, 1946, but it is parked in front of the original terminal building, which was gone by mid-1947. Over the next fifteen years, with 8–10 daily round trips to Vancouver and 3–4 to Seattle, it is estimated that these famously reliable Douglas aircraft completed 150,000 flights in or out of Pat Bay, and carried some 3,200,000 passengers, maintaining a flawless safety record.

Between 1946 and 1948, Victorians had ambivalent feelings about Patricia Bay Airport. Lower Island residents were pleased to have full TCA service, but the long drive to Sidney was still viewed as a major drawback. There was also no guarantee that civil aviation would enjoy a high priority at the postwar air force base.

Two competing remedies were proposed. Spearheaded by enthusiasts who were trying to establish a new Victoria Flying Club, some citizens hoped to revive the idea of a municipal airport at Gordon Head. This might or might not be suitable for commercial aircraft, but it could at least accommodate small private planes. The Gordon Head Memorial Air Park campaign, headed by Victoria's "Lighthouse Philosopher" William Scott, will be reviewed in Chapter 18.

The other school of thought, espoused in 1946 by Transport Minister C.D. Howe and the Ottawa Liberal Government, was to dismiss the notion of an airfield at Gordon Head and spend available funds on a proper Patricia Bay Highway. That cause eventually won the day, though the original East Saanich Road continued to provide the only standard access until the mid-1950s.

Two significant and promising developments occurred in the early postwar years:

On August 1, 1947, RCAF Hangar No. 4 in East Camp, a structure 160 feet long, was opened as the new TCA Administrative Building and Passenger Terminal at Patricia Bay, replacing the tiny converted farmhouse to the north. Refurbished Hangar 4 is seen across the page.

On March 31, 1948, the RCAF flag was officially hauled down in the presence of its last C.O., Group Captain F.R. West. Patricia Bay Airport and its facilities were now taken over by the Department of Transport.

Meanwhile in 1946, two native Victorians, long-time friends and associates, were doing well with TCA.

*TCA Between Ourselves* (May 1946: 5–6). Maurice McGregor Collection

B.C. Archives I-01795

Many Victorians will remember this plain, friendly passenger terminal, East Camp Hangar 4, in use from August 1947 until August 1964, when a new terminal opened on the south side of the airport. The picture above appeared in the *Colonist* on Sunday, December 12, 1948, p. 11.

This TCA terminal was threatened by two major fires soon after it opened. On the night of October 28, 1947, Hangar No. 5 (directly east of 4) was completely destroyed by fire, along with seven Stranraer flying boats. On August 21, 1948, a second fire destroyed the rear section of Hangar No. 4, but the building was saved.

A uniquely Canadian aircraft of the 1950s, the Canadair North Star was a modified Douglas C-54/DC-4, fitted with Rolls-Royce Merlin engines. Fast but noisy, it was widely used by TCA and the RCAF. At right is an RCAF North Star at Pat Bay Airport. Behind it are two TCA DC-3s, and on the left is a Queen Charlotte Airlines Anson.

Photo by Elwood White

Photo by Elwood White

George Maude Collection

The two pictures to the left both show TCA North Stars. In early December 1956, Elwood White took the upper photograph, which depicts an aircraft on the brink of tragedy. Apparently diverted to Victoria because of inclement weather, this is TCA CF-TFC (fleet number 204). Days later, on December 9, it crashed on Mount Slesse near Hope, killing all aboard, including five football players returning from a CFL East–West All-Star game. It was one of the worst disasters in Canadian aviation history.

Happier memories are associated with the lower photo, which shows a North Star air freighter (recognizable by its covered windows). It is a mild and sunny day in late winter, sometime during the mid- to late-1950s. TCA employees are loading boxes of daffodils from the farm of G.A.Vantreight, Jr., for shipment to the eastern Canadian market. Scenes like this have now been annual Victoria rituals for over sixty years.

The photograph is from the collection of George Maude, who is one of these TCA employees. He has identified the equipment as the original belt-loader and tractor at Trans-Canada Air Lines, Patricia Bay.

B.C. Archives I-26681

Resplendent and alone on the tarmac is Trans-Canada Air Lines DC-3 CF-TDR, the same aircraft shown in front of the terminal on the previous page; the two B.C. Archives photographs were likely taken on the same occasion. As Vancouver-bound passengers board the plane, perhaps in early winter 1948, we have a panoramic view towards the wartime buildings of West Camp, the Officers' Quarters and Mess on higher ground near Mills Road, and the original control tower (upper right).

Vaguely familiar waiting rooms from our past evoke strong memories. Given the aura of gleaming emptiness, this looks like an official photograph of the TCA terminal, which opened on August 14, 1947. Signs on the wall point right "COACH FOR VICTORIA" and left "SEATTLE." Access was directly off East Saanich Road; the new Pat Bay Highway would not be built until the 1950s.

When we compare this one simple room with the spacious, handsome passenger terminal recently enlarged at Victoria International Airport, we may marvel at the progress made in just over half a century. The Victoria Airport Authority deserves praise for its role in this achievement.

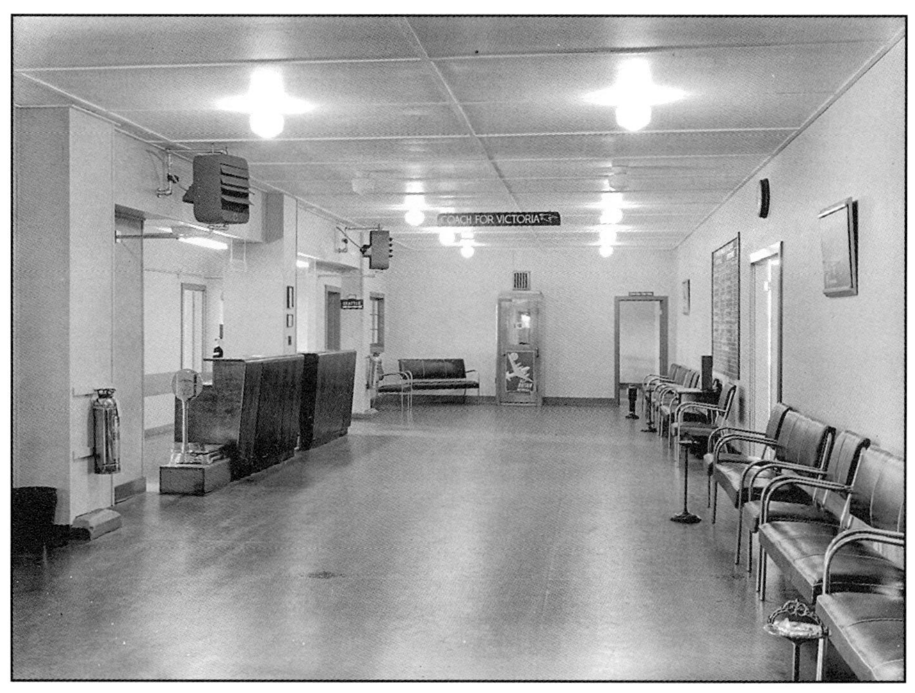


B.C. Archives I-01795

## 18. Victoria Flying Club: Gordon Head Revisited ▪ 1946–1948

Photo by Elwood White

Before scenes like the one above could be realized at Patricia Bay Airport, several years of frustration had to be endured by proponents of a post-war Victoria Flying Club. The Club seemed such an obvious and reasonable idea, but the obvious and reasonable are not always quickly attainable.

There were several problems to be overcome, starting with the right to exist. In September 1946, the would-be Flying Club was informed by the Deputy Minister of Transport that its incorporation had been refused, partly because Victoria had no airport and partly because the city's former Aero Club had failed for lack of support. The *Colonist* of September 11 editorially dismissed these grounds as "Utter Nonsense"— and rightly so. The pre-war Aero Club lost two fields, and Pat Bay Airport did not exist. Now there was a throng of returning veterans. Even when Victoria was much smaller, there had always been avid flyers; one of them, Gordon Cameron, was still active behind the scenes as a Flying Club supporter.

That sample of bureaucratic inanity was soon countered, and the Club had its charter by November. There was still the genuine problem of an *accessible* airport. It was not at all clear that Patricia Bay was going to be made available for civilian use. Anyhow, Pat Bay was 18 miles north of downtown Victoria, and that still seemed too far for practical use. Other cities, especially in the western U.S.A., had municipal airports at their doorstep.

So arose a campaign to revive the old and long defunct Gordon Head Airport of the 1930s. The site of that field to the west of Finnerty Road had now been pre-empted by the Gordon Head Army Camp, but there was a similar flat area in Saanich to the east of Finnerty. During the war this had been a top secret radio listening post (a so-called Y-Station), operated by men and women of the Royal Canadian Navy. Now it was being held by War Assets as a disposable property. What is more, the City of Victoria currently held an option to buy the 57.5-acre parcel for possible use as a small municipal airport.

The idea had its strong boosters and equally strong opponents. The opposition was led by Gordon Head ratepayers, who had fairly obvious reasons for not wanting an airport in their back yard. Many other citizens with no personal interest in private flying saw Patricia Bay Airport as quite adequate for all future aspects of local aviation. This group sided with the opposition.

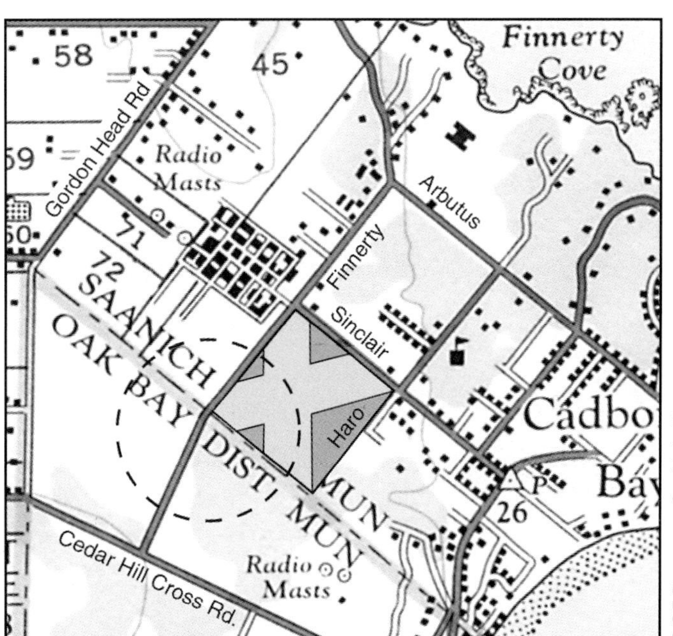


D.O.T. Plan courtesy Saanich Archives

The post-war Gordon Head Airport was across Finnerty Road from its predecessor of the 1930s (page 71), now transformed into Gordon Head Army Camp. This map is the conflation of a Department of Transport runway plan, ca. 1946 (shaded area) and a standard topographical map from the 1950s. Editorial additions include six street names and the location of the future UVic Ring Road.

Photo by Elwood White

Patricia Bay Airport, May 1948: Victoria Flying Club Fly-in, with visiting aircraft from Vancovuer and Chilliwack.

The *Victoria Times* opposed the project, as did the Liberal Government in Ottawa. The Victoria Flying Club was determined, however, and was staunchly supported by the Conservative *Daily Colonist*. Ardent boosters envisioned an expanded municipal airport that would extend across Finnerty and south in Oak Bay to Cedar Hill X Road.

The goal was soon defined as Gordon Head Memorial Air Park, a small 57-acre civic airport for light aircraft only, intended to honour the veterans of World War II. Outspoken leader of this cause was the visionary but unconventional W.A. (Bill) Scott, Victoria's "Lighthouse Philosopher." His proposal was unanimously adopted by the Club at a meeting on September 27, 1946; but the issue was not resolved until December 1948. Meanwhile, the Club gained temporary access to Pat Bay Airport and Hangar 3, thanks partly to the good offices of Robert W. Mayhew, M.P., who was trying to play Solomon with the opposing factions among his Victoria constituents.

In May 1947 the City of Victoria actually bought the Saanich property for $20,000, the sale to be contingent on its eventual use as an airport. In 1948, members of the Flying Club set to work grading a 1,600-foot runway.

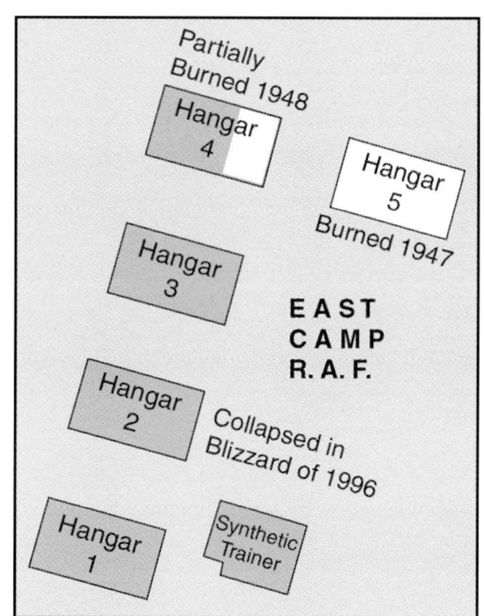


**HANGAR 4**—TCA terminal from 1947 to 1964; since1965 it has housed the Victoria Flying Club.

**HANGAR 3**—at first shared between VFC and Airport Administration (Department of Transport); it was used also by Flying Firemen.

**HANGAR 2**—occupied by Fairey Aviation, and then by Viking Air. It collapsed in Victoria's blizzard of December 1996. Fairey also used the RAF Synthetic Trainer building.

**HANGAR 1**—shared for a while by the Government of British Columbia and Pacific Western Airways, who had a contractual relationship.

Photo by Elwood White

February 1948. Four VFC aircraft stand outside Hangar 3: two ex-RCAF trainers also in top photo, Cessna Crane CF-CSH and Cornell CF-EMV; a Fleet Canuck; and Piper Cub CF-DAR, formerly owned by Dave Duncan.

Perennial treasurer of the Victoria Flying Club, Claude O. Butler (1911–1986) was among the most influential local aviators in the years 1946–1971. Not only did he have a passion and a natural gift for flying, he also gave generously of his time and exceptional business skills. He served as president of the Victoria Flying Club, and also of the Saanich Board of Trade.

In partnership with his friend Joe Howroyd, he established an important community airfield on Keating Crossroad, near his pioneer grandparents' property and the Butler Brothers logging, hauling and construction business in which he played a major role. For this Butler-Howroyd Strip, see Chapter 20.

G.E.M. Mortimore devoted one of his excellent Sunday *Colonist* profiles to Claude Butler, presented as usual with a Mort Graham drawing (see right). He tells the story of how the four Butler brothers built up their company from scratch, by dint of hard work, in the unforgiving climate of the 1930s. Mild and modestly self-effacing Claude, a genuine financial wizard, acted quietly and diplomatically as their leader.

Left with a pronounced limp after severely gashing his leg in a childhood bow-making accident, he nonetheless possessed great vigour and energy. An avid motorcyclist, a practical craftsman, and a fine musician, Claude Butler was clearly a man of many parts. Most impressive is the fact that no one has a bad word to say about Claude, who was considered an exceptionally decent and civilized human being.

Claude Ormond Butler
Drawing by Mort Graham, *Colonist Islander* 1955.09.11

It was a battle to the end with Saanich Ratepayers, a conflict well summarized by Ursula Jupp (1975: 172–174). After one runway had been laboriously graded and groomed, Frank Copley made a single landing in a Fleet Canuck on November 7, 1948. He was then president of the Flying Club; his passenger, Claude Butler, was president of the Saanich Board of Trade. Only one other landing is attested, appropriately enough by Joe Howroyd, who lived within a half mile (Chapter 20).

The rezoning of the land for airport use was challenged by ratepayers and a referendum to test that zoning was put to Saanich voters in December 1948. It was solidly defeated, 3,437 to 1,706—and so died the noble dream of a Gordon Head Memorial Airpark (luckily for UVic).

As early as 1948, the Victoria Flying Club seemed well settled in Hangar 3 at Patricia Bay, and was soon acquiring many new members and a growing number of aircraft. There was no lack of enthusiasm or energy. In contrast to a parallel situation after World War I, not many of the prime movers were returning Air Force veterans; it may be that most of those had done enough flying for a lifetime. Some veterans, like Stranraer pilot T.B. (Bert) Toye and Hurricane pilot J. David Duncan were enlisted as instructors, and others became active members. Still, most of the prominent Flying Club members—Frank Copley, Joe Howroyd, Claude Butler, Morris Kersey, H.B. (Barney) Oldfield, and Dr. Arthur Nash, to name a few—began as enthusiastic novices who yearned to fly, and having learned, never lost their zest for flying.

Space does not permit us to present a history of the Club beyond the first few years, though a number of later photographs will be included. Early presidents not yet mentioned include F.T. Sehl, Fred Marconi, and Myles I. Hague; J.B. (Jack) Taylor was also very active as vice-president and manager in the inaugural year of 1946.

John Howroyd Collection

The Frank Copley family is photographed with Frank's Stinson, CF-EIV.

# 19. Two Commercial Enterprises in the Early Postwar Years

## William Beresford Sylvester & His Victoria-Based B.C. Air Lines

George Williamson Collection

B.C. Air Lines Waco (CF-BBO) at Limekiln Cove, 1946.

There are not many surviving clear photographs of seaplanes at the Stewart Avenue wharf in View Royal. Limekiln Cove or Helmcken Bay is at the Esquimalt end of Helmcken Road. That small bay was Victoria's Canadian Pacific Airlines base from 1942 to 1945. CPA had flown DH Dragon Rapides and Barkley-Grow T8P seaplanes from this wharf, which was still partly extant in 2004.

His local credentials were impeccable: his father, also William B. Sylvester, was born in Victoria in 1874. Young Bill was eighteen—the same age as Maurice McGregor and Claude Butler —at the time of the 1929 stock-market crash. The tough years of the Great Depression helped him develop a shrewd business sense. During those Dirty Thirties, Bill Sylvester was fully engaged in the U-Drive business—on which he held a virtual Victoria monopoly—but he also took a keen interest in aviation.

Bill actually learned to fly at Esquimalt some years before he received his licence, which he earned at Foggins Flying School on Sea Island (site of Vancouver Airport). During the Second World War, he acquired two Luscombes (CF-BPA and CF-BSR), and in 1943 he incorporated the company that he named B.C. Air Lines. He was clearly looking ahead to postwar days: he stored the two machines in a downtown Victoria warehouse until a time when commercial aviation might get into full swing again after the war was over.

B.C. Air Lines was granted the first Air Transport Board charter licence in British Columbia, and only the fourth in all Canada. The budding airline was on its way to becoming the largest home-owned air service in the Province. Eventually it would have sixty employees, twenty of them pilots.

AIR TAXI

B.C. AIRLINE LTD

CF-BBO

B.C. AIR LINES LIMITED

WISH TO ANNOUNCE

That on and after December 1, they will operate from the former C.P.A. Base, foot of Stewart Street, View Royal. The above pictured aircraft will be open for inspection on Sunday and available for sighting trips over the city.

24-HOUR SERVICE

• CHARTER • PLEASURE • EMERGENCY TRIPS

CITY OFFICE — SUSSEX HOTEL

Bill Sylvester's Waco ZQC6 (CF-BBO) was advertised as a Victoria-based air taxi. For its fate, see next page.

Bill Sylvester's B.C. Air Lines started active operations in Esquimalt Harbour. In 1944 Bill had bought a Waco Custom biplane (CF-BBO) from Grant McConachie of Canadian Pacific Airlines; in late 1945 he based this aircraft at the former Canadian Pacific float on Stewart Avenue, in View Royal. From there he ran his first taxi and charter service. An unfortunate accident on September 20, 1946 saw the demise of this seaplane, which stalled and crashed on takeoff from Esquimalt Harbour. The aircraft was demolished, but pilot Stan Berge and three U.S. businessmen aboard the plane survived the crash with various injuries.

B.C. Air Lines Waco CF-BBO is seen above on wheels, after a rebuilding job in Edmonton. At right, it was damaged beyond repair on September 20, 1946.

George Maude Collection

Seen above at Patricia Bay is Luscombe CF-BSR, one of Bill Sylvester's first two aircraft, which he had stored during the later years of World War II behind Sylvester U-Drive in downtown Victoria. At right is that same CF-BSR with Sylvester's chief pilot and right-hand man, George Williamson. Below is yet another B.C. Air Lines Luscombe, CF-FGU, photographed at Pat Bay with a B.C. Government Anson (CF-BCA) in the distance.

George Maude Collection

George Williamson Collection

Realizing that Vancouver offered greater potential, Sylvester soon moved his headquarters to Sea Island, leaving the Victoria operations under the supervision of Stan Scurrah. Management of the Vancouver base became the responsibility of George Williamson. The fleet grew when Bill took over the assets of Westinghouse Airways of Victoria (below), which had operated a Luscombe and three Seabees.

Throughout the early 1950s the company continued to expand under his direction, as he established new bases on Vancouver Island at Campbell River and Port Hardy, and on the Queen Charlotte Islands at Sandspit. B.C. Air Lines also created about a hundred radio-telephone stations in as many logging camps, fishing villages, and small isolated communities along the B.C. coast. These became a lifeline between local residents and the outside world. The number of mercy-flights rose to an average of 500 a year, and in later years people could not speak too highly of the pilots who flew BCAL's small machines.

The wider history of B.C. Air Lines lies outside our scope, though we shall again see Bill Sylvester back in his home town with Victoria Flying Services. In 1956 he sold his first airline to F. Maurice McGregor (once again a Victoria resident) and Paul Tak. By then the BCAL fleet had grown to 25 aircraft.

Many great stories are told about Bill Sylvester, who was a very rugged and colourful character. For further reading, we highly recommend Jack Schofield's *No Numbered Runways* (Winlaw, BC: Sono Nis, 2004), which has a fine chapter on Bill.

## Aubrey Westinghouse: Westinghouse Airways ▪ 1946–1954

1950 Victoria Telephone Directory (Yellow Pages: 2)

For almost a decade (1946–54), brothers Aubrey and Lawrence Westinghouse—grandsons of U.S. inventor and industrialist George Westinghouse—ran a charter air service, first out of Vancouver and then Victoria (their head office). Aubrey was born in Massachusetts in 1913, and learned to fly as a teenager in Arizona. Around 1940 he settled in Saanichton with his wife Connie. After wartime service with the RCAF and RAF in England and North Africa, he and his younger brother Lawrence founded Westinghouse Airways in 1946. Their company flew passengers and merchandise up and down the B.C. coast, in both amphibious and wheel-equipped aircraft.

Westinghouse Airways used a Luscombe and three Seabees, one of which is shown above. In Victoria, Westinghouse seaplanes operated out of the same Stewart Avenue base in View Royal that had earlier been used by Canadian Pacific and B.C. Airlines. In 1954, Westinghouse Airways was sold to B.C. Airlines. Aubrey's son George became a widely experienced commercial pilot and a teacher of aviation at Selkirk College in Castlegar.

Jim McKeachie, RCAF veteran and future CP Air executive, featured Westinghouse Airways in an interesting 1948 article: "Flying Salesmen Pioneer Air Road." (*Times* 1948.08.07: mag. sect. 3)

# 20. Private Air Strips Begin to Flourish in Victoria ▪ 1946–1965

## Joseph Richard Howroyd and his Home-Grown Daffodil Strip

One of southern Vancouver Island's keenest aviation enthusiasts was the late Joe Howroyd. Born in Victoria in 1916, he took an avid interest in early flying events at Lansdowne Airfield, where, as a boy, in order to learn everything possible about aviation, he would help clean and fuel aircraft, doing any chore that would keep him near airplanes. At the age of twelve, he painted in the registration letters G-CATX on B.C. Airways' ill-fated Ford Trimotor. An excellent mechanic, he was always maintaining or modifying machinery. Whether it was a machine on his daffodil bulb farm off Shelbourne Street in Gordon Head, a motorcycle (he was an early member of the Victoria Motorcycle Club), his racing speedboat *Johnny Boy* at Shawnigan Lake, or any airplane—his equipment was kept in meticulous condition.

John Howroyd Collection

Shortly after World War II, Joe took his flying instruction at Victoria Flying Club, receiving his private licence in 1946. Soon after, he acquired his first aircraft, a Cornell (CF-FQT), which he flew from his bulb farm on what is now a Saanich subdivision north of Mortimer Street, enclosed by Shelbourne, MacKenzie (then Ruby), and Gordon Head Road (see map, p. 163). His personal runway evolved into "Daffodil Strip," a small private air field that was the first base for Vancouver Island Helicopters, and was used by other local enthusiasts such as Claude Butler and Frank Copley. Residents of Gordon Head in the 1950s might often see a small plane making an east-to-west landing approach on a path over what later became Campus View Elementary School, before the aircraft disappeared in the direction of Shelbourne Street. Daffodil Strip was in occasional use as late as 1962—the year UVic began construction at Gordon Head—when the fast-growing area became too congested for further flying activities.

Foreseeing the need for a more substantial private airstrip, Joe Howroyd and Claude Butler obtained land in the Keating district of Central Saanich, where in 1955 the Butler-Howroyd Airstrip came into being, its first hangar a converted barn. Over the years a number of new hangars were built, and the field soon became the hub of sport flying in the Greater Victoria area. In its heyday, the Butler-Howroyd Strip, known also as the Keating Strip, provided a base for an average of thirty light aircraft, as well as Unicom facilities, gasoline, and oil. For the next forty years, this operation was a great success story, little known beyond the tightly knit circle of aviation enthusiasts. To the regret of many, the field became gradually diminished in size: it was located on top of one of the region's prime gravel deposits, which was being steadily used and eroded.

The Howroyds were Victoria's flying family, son Johnny receiving his pilot's licence on his sixteenth birthday, becoming Canada's youngest pilot at the time. Mrs. Iris Howroyd also earned her private licence in 1964. Upon selling the Cornell, Joe bought Stinson CF-HUL and later Stinson CF-JYN, in which he had his only mishap. On takeoff from the Butler-Howroyd Strip, the throttle linkage malfunctioned and the machine plunged over a high bank into the telephone wires below. No serious injuries were suffered by Joe or his passengers, but the Stinson was rather badly bent. A pretty little Luscombe CF-HUM, christened *Hummin' Bird*, was Joe's next machine; painted white with red trim, it attracted much attention from admirers. Joe's pride and joy was a Mooney Mk.21, CF-MST, named *Misty*. Trips as far away as the antique Fly-in at Rockford, Illinois, the Indiana 500, and air-meets in California and Nevada saw Joe in attendance, renewing acquaintances with other well-known aviation personalities. When CF-HUM was sold, the Howroyds acquired another Luscombe, CF-SHE. As was the case with all his aircraft, it was restored to mint condition, and won many awards at various meets around the Pacific Northwest.

One of Joe Howroyd's many good qualities was his unflagging help to others, given always without hesitation. Many of the light planes brought to this area came through his purely voluntary efforts. He would arrange trips to distant places in order to appraise and ferry machines to Victoria for new owners. His advice, hard work, and encouragement made him sport aviation's number one booster in this region.

When Bill Sylvester formed Victoria Flying Services in 1959, Joe became a director and shareholder, along with other prominent Victorians. One of his favourite duties was to ferry new aircraft from the Cessna factory to Bill's Cessna agency in this city.

Joe died suddenly on March 10, 1969, a loss felt acutely by all who knew him. The Butler-Howroyd Airstrip then became ably managed by son John, like his father an avid sport-aviation enthusiast.

Joe, Iris, and Johnny Howroyd in Joe's first plane, Cornell CF-FQT about 1950, when young Johnny was five. Victoria Flying Club, Pat Bay Airport.

Above: Ernie Eve and Mrs. Eve with Joe Howroyd at Dog Creek, B.C., where Ernie was Airport Manager. Joe is about to ferry Ernie's Cornell CF-EMV from Dog Creek to Victoria, for new owner H.B. (Barney) Oldfield.

Below: Joe Howroyd visits a wintry Dog Creek in his new Stinson CF-HUL.

**Other Early Airstrips on Southern Vancouver Island**

- Parkridge, between Carey Road and Mann Ave., built by prominent sport flyer Franklin Malahat J. Copley in 1952.
- Twin Totems Farm, off Island View and Puckle Road in Central Saanich, built by Ron Page and later owned by Frank Bernard.
- Arden Airstrip just past Sooke, built by Butler Brothers close to their logging operations.
- Jackson Strip at Cherry Point, near Cowichan Bay.
- The Frank Norrie Strip, near Dougan's Lake at Cobble Hill.

Above are two views of Joe Howroyd's Stinson CF-HUL at Daffodil Strip. To the right, a group of schoolchildren admire Ted Henson's helicopter, CF-IDX, which was kept originally at Daffodil Strip. See Chapter 21.

At the bottom of the page are Joe Howroyd's two Luscombes: CF-HUM (*Hummin' Bird*) and CF-SHE (shown with Joe), with which he won a trophy not long before his death in 1969. CF-SHE is still owned in 2004 by Johnny Howroyd.

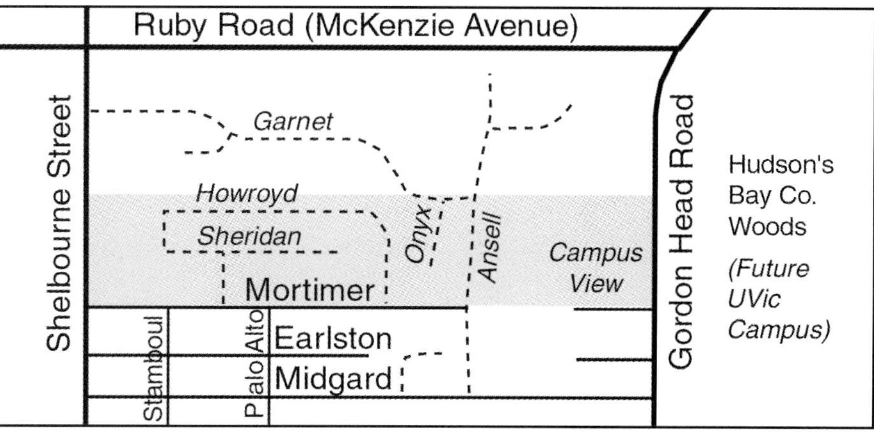


Daffodil Strip is the shaded part on the map; the entire area between Mortimer and Ruby Road was undeveloped. The 60-acre Howroyd bulb farm at 1710 Mortimer ran from Gordon Head Road west to Shelbourne. The main "runway" was the strip now enclosed by Onyx, Howroyd, and Sheridan. In 1956, ten acres were sold to the Greater Victoria School Board for what became Campus View Elementary; another parcel along Shelbourne was sold in 1957–58.

All photographs on this page courtesy John Howroyd

## Butler-Howroyd (Keating) Strip in Central Saanich

John Howroyd Collection

By the early 1990s (above) Butler-Howroyd airfield on Keating X Road was reduced to an ever-shrinking plateau threatened by encroaching gravel pits—but the hangars testify that it was still well used. Back in 1955 (right), the new strip was surrounded by farmland. In the upper photo we look northwest over the strip toward Brentwood Village, with Keating X Road meeting West Saanich Road at left centre. In the 1955 picture, taken from Joe Howroyd's Stinson, we look southeast to Bear Hill, with Little Saanich Mountain on the right. Keating X Road is barely visible in the dark foreground; the high-perched airfield has always been beyond the line of vision for motorists driving that thoroughfare.

John Howroyd Collection

John Howroyd Collection

The Howroyd-Butler partnership was based on mutual respect and warm friendship. Here is Claude Butler (left) with young Johnny Howroyd on Joe Howroyd's bulb farm. They are standing beside Claude's Cessna 140 (CF-ECP), on the day in May 1953 when he sold that aircraft to Barney Oldfield. Until Claude Butler and Joe Howroyd developed the airfield at Keating X Road, Claude regularly flew from Daffodil Strip.

## SKYWAY AIR SERVICES, LANGLEY: FERTILIZER CROP-DUSTING FROM KEATING STRIP (1964)

Three views of Boeing Stearman CF-FBV along with Stearman CF-DFC, during a session of fertilizer spraying from Butler-Howroyd Strip on May 2, 1964.

(Above) Two Stinsons at Keating Strip in April 1965. (Below) Stinson CF-EIM landing at Keating on April 2, 1967.

All photos by Elwood White

# 21. The Origins of Vancouver Island Helicopters ▪ 1953–1957

Though the story of Vancouver Island Helicopters has been well told by Peter Corley-Smith and David N. Parker (1998: 62–75), our book must include a brief tribute to Ted Henson, the resourceful Victorian who founded that company in 1954, after receiving a remarkable boost from Bill Boeing Jr. Obviously impressed by the ambitious young Canadian, Boeing financed Henson's helicopter instruction and the down payment of $18,000 on a Bell 47-G, registered in Canada as CF-IDX.

Fellow Flying Club member Joe Howroyd also gave Ted Henson strong support, letting him store the helicopter at Daffodil Strip. That is where it hovers in the top picture: (L–R) Joe Howroyd, Ted Henson, and Boeing pilot Gerry Garbell. In the second picture, the helicopter is alongside Joe's Stinson, CF-HUL.

Officially licensed by 1955, Ted moved CF-IDX to a tiny hangar at Pat Bay Airport, prepared to tackle any commercial job that came his way. Progress was very encouraging; by 1957, when the company's fleet had grown to six aircraft, Vancouver Island Helicopters was enjoying considerable success. Elwood White's Pat Bay Airport slide of CF-IDX, lower right, is dated February 1957.

Tragically, Ted Henson was killed on October 17, 1957, when CF-IDX crashed into Tatlayoko Lake on the B.C. mainland. Despite this cruel setback, his company was saved and successfully developed by his wife Lynn. The next three years led to impressive growth. In 1960, Lynn Henson married Alf Stringer, Operations Manager of Okanagan Helicopters. In 1963, Alf Stringer left the opposition to become president of Vancouver Island Helicopters.

John Howroyd Collection

John Howroyd Collection

Photo by Elwood White

# PART SEVEN

# VICTORIA INTERNATIONAL AIRPORT

## 1959–1971

## 22. New Airport, New Terminal, New Air Canada ▪ 1959–1971

Victoria International Airport. How much more impressive that sounds than the old Pat Bay; yet the name change—initiated by Victoria Mayor Percy B. Scurrah—became effective on May 22, 1959 with so little pomp and publicity that one wonders if residents of Greater Victoria even noticed. Still, it was an obvious fact that Victoria was now a major air terminal on the national map, always in the top ten in terms of passenger traffic. In 1959, the "international" aspect was limited mainly to Victoria–Seattle flights, but that situation would change with the advent of jet travel in the 1960s and 1970s.

Several memorable moments punctuate the years 1959–1971. First was the transition in January 1961 from the faithful, familiar DC-3 to the new 48-passenger Vickers Viscount, a 315 mph turbo-prop, another aircraft that soon won a legion of local admirers. (TCA was already flying DC-8 jets on the national routes, but Victoria passengers have normally changed planes in Vancouver.)

Memorable moment number two was the opening in August 1964 of the new passenger terminal at the south end of the field, an admirable small building that would be altered out of all recognition over the next forty years.

A third milestone was the change in name from Trans-Canada Air Lines to Air Canada, effective on January 1, 1965. This was almost as controversial as Canada's new Maple Leaf flag—for some of the same reasons. On this issue, president Gordon McGregor and the Lester Pearson Government were in full agreement. Like the Maple Leaf flag, the name Air Canada soon won over most opponents, and within a few years it had become another national icon.

Runway 08-26 was extended in 1961 to 6,000 feet, and in 1971 to 7,000 feet (to accommodate DC-8 jets).

In addition to its role in civil aviation, the airport had gained a new military presence: the Navy. In this last period of our century, energized by Fairey Aviation, it also flourished as a base for major industrial development.

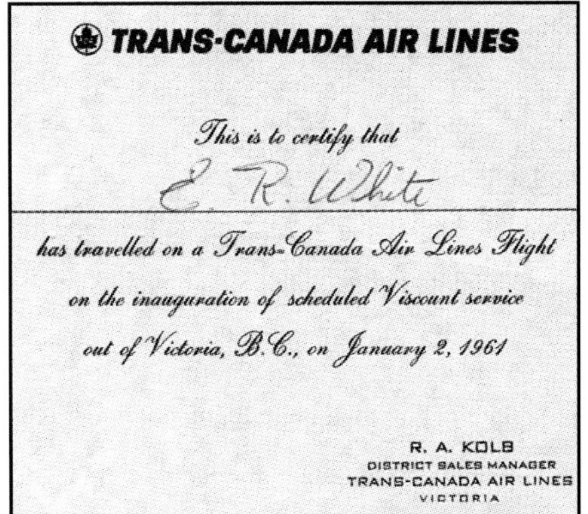

TRANS-CANADA AIR LINES

*This is to certify that*

E. R. White

*has travelled on a Trans-Canada Air Lines Flight on the inauguration of scheduled Viscount service out of Victoria, B.C., on January 2, 1961*

R. A. KOLB
DISTRICT SALES MANAGER
TRANS-CANADA AIR LINES
VICTORIA

On January 2, 1961, there were five courtesy flights—three by the new Viscounts, and two by the old DC-3s. The first scheduled Viscount flight, No. 145 from Vancouver, arrived at 1:55 p.m.

Ten years later, Viscounts were still serving Victoria International Airport, but they now bore the legend AIR CANADA, and had a more fashionable colour scheme. The aircraft at right is CF-THG, photographed in March 1971. Viscounts flew an estimated three million passengers between Victoria and Vancouver from January 1961 to April 1973.

Courtesy *Victoria Times Colonist*

TCA Viscount CF-TGR is the star attraction at Viscount Day, January 2, 1961. A crowd of 3,000 attended, including the now retired Donald R. MacLaren.

Photo by Elwood White

Photo by Elwood White

This shot was taken from a helicopter in May 1971. We see an Air Canada Viscount and a Pacific Western Convair at Victoria International Airport's new terminal. The parking lot looks full.

The new terminal building (right), designed by Wade & Stockdill, was opened at Victoria International Airport on August 14, 1964 (*Colonist* 1964.08.15:16). This is the south facade. Originally, the control tower was to be part of the terminal, but it was decided that this would place it too far from the main runway 08-26.

After three major enlargements in 1974, 1984, and 2003, this original terminal building looks very different in the 21st century.

Courtesy *Victoria Times Colonist*

W.W. Taylor Collection

On March 23, 1947, Bill Taylor photographed the original old control tower at Patricia Bay Airport. For its northern location, see page 154.

Photo by Elwood White

## THE OLD AND THE NEW IN CONTROL TOWERS

In July 1970, a Trans-Canada group flight was staged to commemorate the 50th anniversary of the first Trans-Canada flight in 1920. Most of the aircraft seen here at Pat Bay had just flown across Canada. We get merely a glimpse of the modern control tower at Victoria International, built in the mid-1960s between Hangars 3 and 4; for more complete views of the new tower, see pp. 170 and 182. As the sign on Hangar 4 indicates, the Victoria Flying Club was now in this building (since 1965); a contrasting photo is on page 155.

# 23. The Royal Canadian Navy at Patricia Bay ▪ 1946–1974

In the mid-1960s, four visiting Grumman Trackers from Shearwater, Nova Scotia, assume a "Stand Easy" position. Far across the tarmac, in East Camp, we see the recently completed control tower between Hangars 3 and 4.

First showing in the skies over Victoria of the Royal Canadian Navy's Air Arm occurred when HMCS *Warrior*, en route from Halifax, arrived off Beacon Hill Park early in December 1946. Twelve Fairey Firefly aircraft were flown off the carrier prior to the ship's entry into Esquimalt Harbour. Using the RCAF facilities at Patricia Bay, No. 825 Squadron continued training exercises in this area. A sad loss to the squadron was suffered on January 31, 1947, when Lt. Cdr. O.W. Tattersall, DSC, RN, Commanding Officer of 825, went down with his crew off Portland Island, just four miles northeast of Patricia Bay Airport (Kealy and Russell 1967: 44–45).

The Navy's first already established squadron to be based on the west coast was a detachment of VS 880 Maritime Surveillance Squadron from Shearwater, Nova Scotia, assigned to Patricia Bay in the spring of 1954 to participate in torpedo-running trials. Its presence stirred some controversy. The *Victoria Times* of April 22, 1954 contains a front-page complaint from TCA pilots that Navy aircraft were flying low near the western approach to the main runway. On its second front page, the same newspaper describes (with vivid illustrations) an incident near the Victoria Flying Club Hangar, in which a Navy 8-ton Grumman Avenger "chewed" a 1,400-pound Fleet Canuck in a taxiing collision. VFC's chief instructor David Filby and student pilot David Turnbull were lucky to escape serious injury.

Approval from Ottawa was given to the creation of a permanently based utility squadron to be stationed at Patricia Bay, using the former RCAF facilities. Thus VU-33 Utility Air Squadron was formed on November 1, 1954, with two AS-3 Grumman Avengers from 880's detachment. Its first C.O. was Lt. Cdr. Douglas J. Fisher, CD, RCN, who had 23 men under his command.

In October 1954, that same year, a visit to the west coast by HMCS *Magnificent* brought additional squadrons, on a temporary basis. While approaching harbour, she flew off VS 881 squadron, composed of Grumman Avengers, and VF 871 squadron, flying Hawker Sea Furies. Both paid a short visit to Patricia Bay, before moving on to Vancouver. Once again, many air exercises with ships of the fleet were carried out by these squadrons, while in west coast waters.

In this DND publicity photo dated June 22, 1951, RCN Grumman Avenger 86233 AB:B is seen flying above the carrier HMCS *Magnificent*, its "sea-going airport."

VU-33 then played an important role in Pacific Command, including exercises with ships and establishments, target towing, photography, search and rescue, and general maintenance flying. It gained a third Grumman Avenger in June 1955, and flew Piasecki HUP-3 helicopters for almost six years, from July 1958 through February 1964 (see photographs below).

Its Grumman Avengers were retired in 1960, and replaced by Canadian-built de Havilland Grumman CS2F-1 Trackers. The first of the Canadair T-33 "Silver Star" jet trainers also arrived in 1960. For the next fifteen years, VU-33 was a small but highly versatile squadron whose publicized achievements ranged from rescuing a seriously injured 13-year-old boy from a ledge on Mt. Finlayson (April 15, 1963) to apprehending a Japanese trawler illegally fishing in Canadian waters (the *Koya Maru II*, February 1974). Periodic visits by elements of VS-880 from Shearwater made the RCN facilities at Patricia Bay a busy place during joint exercises.

After twenty good years at Patricia Bay, VU-33 Utility Squadron lowered the flag on July 30, 1974. The next day, with six aircraft (3 Trackers and 3 T-33s), the Squadron moved to Canadian Forces Base, Comox.

DND Photo No. 46663

Above is Piasecki HUP-3 helicopter No. 245, an early arrival in 1958; it was previously numbered 945 (CMFT 1983: 16). Search and rescue became a specialty. Below is Piasecki 622 in a practice lift; VU-33 also flew No. 621, now restored in the CMFT Museum at Langley.

On December 1, 1953, even before VU-33 was organized, VC-922 Naval Air Squadron had been formed as tender to HMCS *Malahat*, the Navy's west coast Reserve, using two Harvard trainers from Cadet Flying Unit 1. These aircraft had come to Pat Bay from Shearwater in the summer of 1952 to provide a means of air familiarization for cadets of Royal Roads Military College.

A busy program of qualifying and re-qualifying pilots kept the small squadron busy with their Harvards and Expeditors; there was also some shared use of VU-33 aircraft. This VC-922 squadron won the Naval Reserve Flying Trophy on three occasions. By 1961, VC-922 had four Beech Expeditors in service. Lt. Cdr. D.J. (Dave) Slader was the Commanding Officer, and Dave Filby of Victoria Flying Club was the air engineer in charge of maintenance. Fairey Aviation Company performed all the major overhauls.

DND Photo No. 73009

Photo by Elwood White

In February 1956, a Beech 18 (C-45) Expeditor was acquired by VU-33 Utility Air Squadron. By 1961, VC-922 Naval Air Squadron (see left) had four aircraft of this type, used mainly for light transport duties.

Grumman CS2F-1Trackers replaced the Avengers in 1960. Tracker 534 is seen, right, with an RCAF DC-3. Nicknamed the "Stoof," the Tracker was a very effective aircraft for anti-submarine surveillance, as it was equipped with state-of-the-art magnetic and electronic detection equipment. It operated efficiently, needing a crew of only four.

DND Photo 73011

An incident that reflected well on emergency services at Victoria International Airport occurred on November 27, 1968, when VU-33 Tracker No. 528 reported that its landing gear would not deploy. The crippled aircraft made an almost perfect belly landing on a foam-covered runway, drawing a loud round of applause from the anxious crew of emergency attendants.

DND Photo E-79336

Known in the U.S. as the "T-Bird" or "Shooting Star," the Canadair version of Lockheed T-33 was dubbed the "Silver Star." VU-33 Squadron acquired three in 1960; designed as trainer aircraft, they were used mainly for naval support in target towing and gunnery alignment. These T-33s were the first jet planes seen regularly in the skies over Victoria.

In two excellent Department of National Defence photographs, T-33 No. 488 is viewed banking (above) as it soars over Esquimalt Harbour and skimming low (right) over H.M.C. Dockyard.

T-33 No. 465 appears on page 184.

DND Photo E-79340

# 24. British Columbia Government and Pacific Western Airlines

The British Columbia Government's first reliance on aviation began with contract services in the 1920s, when Pacific Airways of Vancouver was engaged to conduct Forestry patrols, using Curtiss H5-2L flying boats. Out of this practice grew the Aerial Survey Branch of the Department of Lands and Forests. A Waco seaplane was chartered in 1936 for some months each year, and this aircraft conducted aerial surveys around the Province until the outbreak of war in 1939.

In order to resume this type of mapping work after World War II, the Provincial Government purchased a surplus Avro Anson (CF-EZI) in 1947, and leased another. In 1949 yet another Anson (CF-EZN) was purchased for the Aerial Survey Branch; at the same time the Department of Public Works acquired its first aircraft, Anson CF-BCA. Frank Gilbert was now chief pilot of B.C.'s Air Division.

B.C. (0) 168

Above: War Surplus Avro Anson CF-EZI, purchased by British Columbia's Aerial Survey Branch in 1947. Below: Grumman Widgeon CF-GPJ—Gaglardi's Pride and Joy.

As the needs of the Government grew, it acquired a de Havilland Beaver (CF-FHF), an amphibious machine usually stationed at Kamloops. A Grumman Widgeon (CF-GPJ) was added in 1953; extensively updated, it became a favourite means of transport for Government officials. Because Social Credit Highways Minister P.A. Gaglardi—widely known as "Flying Phil"—soon became B.C.'s most conspicuouly air-minded politician, the Widgeon's registration letters (GPJ) became identified as "Gaglardi's Pride and Joy." Although regarded by critics as an expensive luxury, this aircraft did allow the dynamic Minister to keep tabs on his department's vast highway improvement program.

Photo by Elwood White

In January 1956 it was decided that Pacific Western Airlines of Vancouver should operate B.C.'s Air Division, leasing all aircraft and personnel to the Government as required. Soon PWA was running a large facility at Pat Bay as an operations and maintenance base for Government aircraft, as well as an overhaul base for its own machines.

Using a DC-3, PWA now inaugurated a triangle service on the Victoria–Nanaimo–Vancouver route; but this proved uneconomical and was soon discontinued. Before long the partnership between PWA and the Province was dissolved, and the Government took over its own operation as before.

Bert Toye Collection

Beech 18 Expeditor (CF-BCB) for Aerial surveys.

Photo by Elwood White

Left: Avro Anson CF-BCA: in 1949, it became the first flagship for British Columbia's Department of Public Works. On May 24, 1962, it crashed at Pat Bay, killing both men aboard.

Below: A veteran RCAF and PWA flying boat captain, and a former Victoria Flying Club instructor, T.B. (Bert) Toye joined the Government service in November 1955. Under his responsible leadership as Chief Pilot, B.C.'s Air Division soon earned an unexcelled reputation.

When Ansons CF-EZN and CF-EZI were retired in 1960 and 1963, two ex-RCAF Beech 18 Expeditors, CF-BCE and CF-BCB (page 173), were bought and modified in the Government shops for aerial survey duty. In summer months, a D.H. Otter (CF-BCG, below) was used for topographic survey work.

On May 24, 1962, the original Department of Public Works aircraft, Anson CF-BCA, crashed on takeoff, killing both men aboard—pilot D. Roy McLeod, 43, Assistant Deputy Minister of Highways; and passenger Angelo F. Provenzano, 46, Highways engineer for Esquimalt District. It would be Patricia Bay's only on-field fatal accident during its entire non-military existence. This sad event drew some criticism from the local press (a *Colonist* headline identified the Anson as an "ancient craft"), and there was implied concern about the wisdom of allowing senior civil servants to be also part-time pilots of government aircraft. Mr. Gaglardi staunchly defended his department and its somewhat unorthodox aviation practices.

Bert Toye Collection

In June 1966 a state-of-the-art Lear jet (CF-BCJ) was acquired on a lease-purchase arrangement. It was meant to serve as a fast VIP transport, but the political furore that ensued caused the Government to terminate the lease in May 1968.

A Department of Highways press release, dated March 7, 1969, provides a detailed and very useful overview of the B.C. Government's air services at the end of the 1960s. The special occasion was the first test flight of an amphibian Grumman Goose (CF-BCI, next page), newly converted by McKinnon Enterprises of Oregon to become powered by two Pratt and Whitney 550 SHP (shaft horse-power) turbine engines. The test flight on March 7 was piloted by Bert Toye, assisted by designer and builder Angus McKinnon.

Photo by Elwood White

B.C. Government's D.H. Otter CF-BCG: Patricia Bay, April 1971.

Photo by Elwood White

Victoria International Airport, March 1971. B.C. Government amphibian Turbo-Goose CF-BCI had been modified in early 1969. On June 24, 1971, engine failure caused a forced landing short of the runway. Pilot Ronald Page was seriously injured, and the plane was badly damaged (*Colonist* 1971.06.25: 9).

The original piston-engine Goose had cruised at 130 m.p.h., whereas the modified aircraft now had a top speed of about 210 m.p.h. This project had required the department to trade in its 27-year-old Grumman Widgeon CF-GPJ—"Gaglardi's Pride and Joy"—and to provide additional funding of about $187,000.

In 1967, a Beech 18 Expeditor (CF-BCF) had been similarly converted to turbine power by PacAero Engineering Corporation of Santa Monica, fitted at Pat Bay with a tricycle landing gear and a single-fin tail assembly. This PacAero Tradewind (photo upper right) is on display in the British Columbia Aviation Museum at Patricia Bay.

Bert Toye Collection

Formerly a Beech 18 Expeditor, but now reborn as a PacAero Tradewind, CF-BCF became another favourite of "Flying Phil" Gaglardi. This is the same aircraft, partly refurbished, on display in the B.C. Aviation Museum. The remodelled aircraft above can be compared with a standard Beech 18 Expeditor—a wartime workhorse—such as B.C. Government's CF-BCC, below.

Photo by Elwood White

The 1969 Department of Highways press release noted that the Government's transportation section, comprising eight aircraft under Department of Highways management, was located in No. 1 Hangar at the Victoria International Airport. Eight highly qualified full-time pilots were under the supervision of Chief Pilot T.B. Toye, with training carried out by Captain Ron Page. A well-equipped mechanical shop handled engine overhauls, aircraft modifications, and all regular maintenance and servicing; this shop was staffed by seven licensed engineers under supervisor Hugh Thomas. There was also a radio and instrument shop, stockroom, dispatcher's office, and waiting room.

It was emphasized that less than half the total flying time had been logged by Department of Highways officials, as the service was used by all government departments: "Ministers, Deputy Ministers and other senior personnel have found it invaluable for visits to branches and field operations of their Departments." The government fleet had logged more than half a million miles in 1967, most of it within B.C. There was no doubt that the Province of British Columbia had whole-heartedly taken to the air.

# 25. Fairey Aviation: Martin Mars Water Bombers ▪ 1955–1967

Photo by Elwood White

Summer 1959. In front of the main Fairey hangar at Victoria International Airport are two ex-Navy Grumman Avengers, awaiting conversion to water bombers. They are dwarfed by the newly arrived *Marianas Mars* flying boat.

In April 1955, Fairey Aviation Company of Canada came to Patricia Bay. Having developed a large plant in Atlantic Canada engaged mainly in military work, the company saw a golden opportunity to set up a similar repair base in the west, primarily to handle Navy and RCAF work. The Patricia Bay plant, housed in two East Camp hangars, was established under the management of Dennis Howell.

As this was the only facility of its type between Alaska and San Francisco, it attracted many orders from Alaskan companies, which began to send their work for both conversion and overhaul.

The most notable feat of this company was its contribution to water-bomber or fire-bomber technology. Starting with surplus Navy Grumman Avengers, Fairey converted twenty-one of these aircraft—sixteen for Skyways Air Services of Langley, three for Air Spray in Alberta, and two for Johnson Flying Services in Montana.

Because these were wheel-equipped aircraft, the need to land and fill the tanks was a time-consuming job when speed and a fast turnaround were essential in fighting a forest fire. Fairey's engineers wondered if it would not be possible to scoop up water with a probe while skimming along the surface of the water. An old Junkers W33 seaplane (CF-AQB), with its door off and a length of aluminum pipe fastened between its pontoons, would drone up and down the waters of Patricia Bay, as engineers sought to contrive the most suitable procedure.

The result was the first water-pickup fire bombers in British Columbia: Junkers CF-AQB and CF-AQW, which began to operate for Skyways Air Services in 1959.

Skyway Air Services Grumman Avenger CF- KCJ.

Photo by Elwood White

Photo by Elwood White

Left: The former Pacific Western Airlines Junkers W33 (CF-AQB) at Victoria International Airport in 1959. At right, the same Junkers stands outside the Fairey hangar, alongside the *Marianas Mars* flying boat, the future CF-LYJ.

Biggest and most spectacular of the jobs that Fairey Aviation tackled at Pat Bay was the conversion of three Martin Mars JRM flying boats. Six of these huge machines had been built for the U.S. Navy's California-to-Hawaii run, hauling freight and personnel between 1946 and 1956. Although the first was destroyed in a crash before delivery to the Navy, the others made hundreds of trips, piling up over 75,000 hours.

Named for Pacific Islands (Philippine, Marianas, Marshall, Hawaii, and Caroline), these Mars aircraft are the world's largest operational flying boats. They are equipped with four 18-cylinder Wright Cyclone engines, each with a takeoff potential of 2,500 horse power. With a wingspan of 200 feet, an overall length of 120 feet, and a height of 48 feet, they are truly an inspiring sight to see, either in the air or on the ground.

The Mars acquisition was brought about by Dan McIvor and Doug McFayden, employees of the MacMillan Bloedel Company. Realizing the need for a large-capacity flying boat, they heard by chance that the Mars machines had been retired by the U.S. Navy at Alameda, California. One, the *Marshall Mars*, had been destroyed by fire near Honolulu on April 5, 1950, but there were still four left—now about to be scrapped.

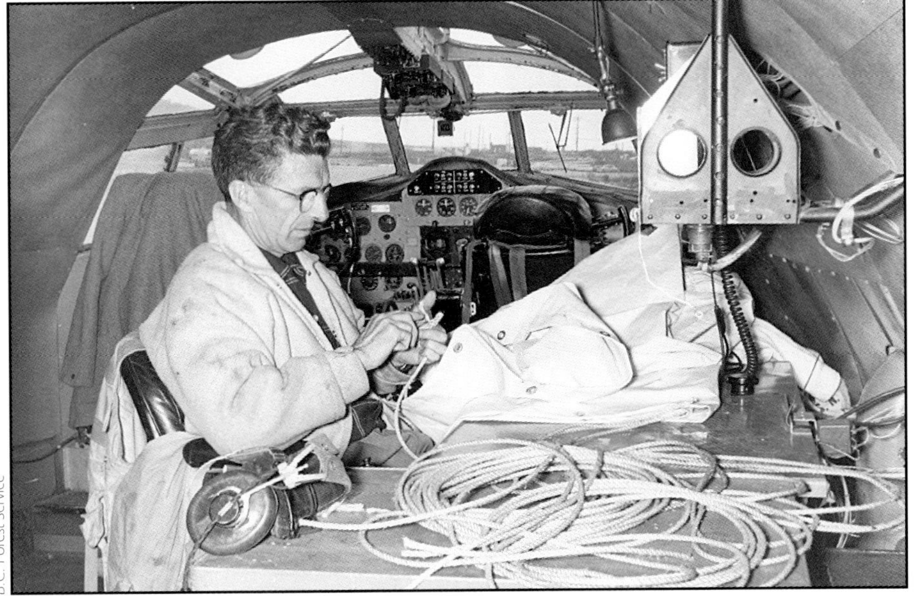
B.C. Forest Service

Seated in the spacious cockpit of a Martin Mars flying boat, Dan McIvor works on solving the technical and logistic challenges associated with the collection and dumping of water on forest fires.

Dan McIvor, MacMillan Bloedel's chief pilot, deserves great credit for his prominent role in the visionary purchase of four former U.S. Navy Martin Mars flying boats, and then his brilliant and highly successful system for equipping them to fight fires. McIvor's exploits in acquiring the four aircraft have been well recounted by A.M. Feast (1990) and Jack Schofield (2004).

For having "revolutionized" the technique of forest fire control, Daniel E. McIvor was invested as a Member of the Order of Canada on April 6, 2004.

B.C. Forest Service 12467

B.C. Forest Service 12464

These B.C. Forest Service photos are dated May 8, 1959, several months before the purchase of the Martin Mars aircraft. At Great Central Lake (left), Dan McIvor and his MacMillian Bloedel co-pilot are planning strategy for dropping water from the MacBlo Grumman Goose named *Dryad*. At right, *Dryad* releases its two 50-gallon tanks.

Dan McIvor's daring plan to buy the four remaining flying boats received swift official support. A consortium of six of the Province's largest timber companies together formed Forest Industries Flying Tankers Ltd., to purchase, convert, and operate the Mars as fire bombers for their own use. (In September 2001, the name was changed to Flying Tankers Inc., which still operates two of these venerable giants in the 21st century).

For $100,000 the consortium acquired the four Martin Mars aircraft—and 93 tons of spare parts! The planes were flown to Pat Bay in August–September 1959, and the conversion contract was let to Fairey Aviation. Each flying boat was allocated a Canadian registration, which became effective when the aircraft entered operation:

| | | |
|---|---|---|
| CF-LYJ | *Marianas Mars* | 1960 |
| CF-LYK | *Philippine Mars* | 1962 |
| CF-LYL | *Hawaii Mars* | 1965 |
| [CF-LYM] | *Caroline Mars* | *1962 |

*damaged beyond repair in Hurricane Frieda; never converted for service.

Courtesy *Times Colonist*

First of the four giant aircraft to be converted for Forest Industries Flying Tankers was *Marianas Mars* (CF-LYJ), left. Project engineers installed a huge, four-compartment, fibreglassed fir plywood tank, 9 feet long and 12 feet wide. This tank was capable of holding 6,000 gallons of water, picked up by two retractable probes located behind the step, enabling the tank to be filled in 22 seconds while taxiing at 70 knots. The water was dumped through four side doors specially built for the purpose.

The complex conversions were made possible by the remarkable wealth of human talent then available at Victoria International Airport. Dan McIvor, still employed by MacMillan-Bloedel, co-ordinated the operation. The Consortium's very first employee as director of maintenance was a man who had begun his local career in 1956 as chief of maintenance for the Victoria Flying Club: now 38 in 1959, Norwegian-born Nils Christensen was a rising star who had proven his technical prowess already in the expert conversion projects carried out at Patricia Bay by PacAero Engineering Corporation and McKinnon Enterprises (Chapter 24). He would serve from 1959 to 1965 as flight engineer on the Mars Water Bombers, alongside engineers and mechanics on loan from Fairey Aviation. Alex Davidson, Fairey's chief test pilot, flew many trial flights with Dan McIvor in order to get the "bugs" out of the system. The scale of the project required a complex variety of creative and technical skills.

Nils Christensen Collection

This historic photograph was taken on the flight deck of *Marianas Mars*. It brings together six men who made major contributions to the conversion and eventual success of the Martin Mars water bombers, which have proven their effectiveness in over forty years of distinguished service. Standing (L–R): pilot Alex Davidson, project head Dan McIvor, engineer Angus Kieler, pilot William G. (Bud) Richmond. Seated: flight engineers Nils Christensen and John V. (Jack) Edwards. Bud Richmond and Jack Edwards were tragically killed in the 1961 crash of this aircraft, and Alex Davidson in the 1967 crash of Flying Firemen Canso CF-FFX (p.182).

March 28, 1960: at Sidney, *Marianas Mars* CF-LYJ staged a spectacular demonstration drop (left). Unlike its two more familiar successors, *Philippine Mars* and *Hawaii Mars*, CF-LYJ had not received a brilliant red-and-white colour treatment, but still wore its original dark blue livery. On July 4, 1960, the first official firefighting mission was attempted, but engine trouble forced the run to be aborted. After the problem was resolved, the *Marianas Mars* flew many very successful missions that year and into the next.

**The Daily Colonist.**

*Vancouver Island's Leading Newspaper Since 1858*

*Island Forecast: Sunny Weekend*
(Details on Page 2)

VICTORIA, BRITISH COLUMBIA, SATURDAY, JUNE 24, 1961 * 10 CENTS DAILY 15 CENTS SUNDAY 26 PAGES

# GIANT FIRE-BOMBER SLAMS INTO HILLSIDE, KILLING FOUR

No sooner had the *Marianas Mars* demonstrated the success of the Martin Mars project when tragedy struck on June 23, 1961. While fighting a fire near Northwest Bay on Vancouver Island, CF-LYJ struck some trees and crashed, killing captain W.G. (Bud) Richmond of Sidney, co-pilot Wallace J. Wiggins of Alberni, flight engineer Jack Edwards of Victoria, and flight mechanic Robert A. Morin of Richmond.

Although for some reason the crew had not released the 23-ton load of water, the cause of the disaster has remained uncertain. Bud Richmond had been a wartime RCAF pilot, a flying-boat veteran with a million miles of experience and an award for 10,000 accident-free flying hours.

This calamity might have brought an end to the bold Martin Mars conversion program; however, the Forest Industries Flying Tankers Ltd. consortium affirmed its support for the project, and decided to move ahead.

Photo by Elwood White

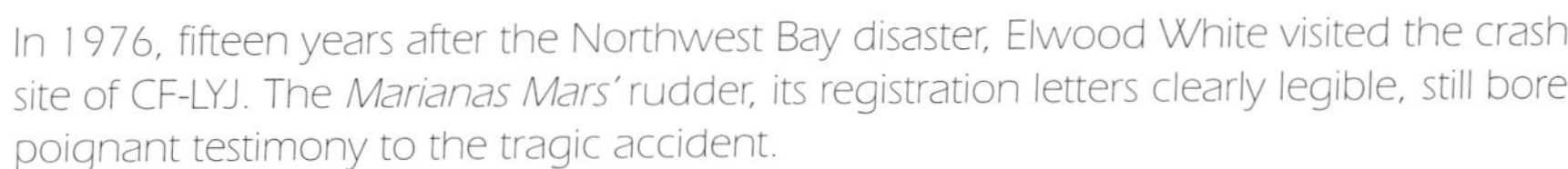

In 1976, fifteen years after the Northwest Bay disaster, Elwood White visited the crash site of CF-LYJ. The *Marianas Mars'* rudder, its registration letters clearly legible, still bore poignant testimony to the tragic accident.

Photo by Elwood White

*Philippine Mars* (CF-LYK) before conversion (above) and after (below).

Photo by Elwood White

The second machine to be converted was *Philippine Mars* (CF-LYK), which was ready for the 1962 season, in a dazzling white and red paint job. By this time, a Cessna 195 was also used on missions as a "Bird Dog" spotter, thus relieving Mars pilots of the risky and onerous task of surveying runs and reconnoitering the fire zone.

The success so far shown by these "gentle giants," as they were sometimes called, led to the third conversion, *Hawaii Mars* (CF-LYL). Here the Fairey engineers devised a different system, by which the load was to be carried within a section of the plane's gas tank. Other technical improvements were also devised, such as a set of doors to be placed in the reinforced bottom skin, thus allowing a much more concentrated drop over a fire. These innovations meant that LYL was not ready for action until the 1965 season.

The last Mars was never to see action, as a result of Hurricane Frieda, a freakish storm that blasted the entire Pacific Northwest on the weekend of October 12–13, 1962. *Caroline Mars* (CF-LYM) literally took flight while securely tied down on the north side of the airport. The incident inflicted serious damage, leaving the machine beyond economical repair. She lay on the field until 1967, when she was scrapped by Victoria's Capital Iron and Metals.

Jack Schofield Collection

Photo by Elwood White

Above are three pictures of *Caroline Mars* (CF-LYM) under very different circumstances. We see her in her U.S. Navy heyday, during a visit to Quonset Point, Rhode Island, in June 1949. Destined never to be converted for water-bomber duty, she had a lingering, undignified end. The left photo is dated 1966. After Hurricane Frieda in October 1962, she lay neglected until December 1967, when she was scrapped (right) by Victoria's Capital Iron and Metals.

Photo by Elwood White

During the peak years of its Patricia Bay operations, in 1960–61, Fairey Aviation had some 200 employees, the majority on cargo conversion work for Alaska-based airlines. To bolster its western Canada branch, the Company set up a printing-press manufacturing section, a plant managed by Ken Hibbert, with Ron Evans as Shop Foreman. But the jet age was to be increasingly influential, as more and more airlines turned to that type of power. When the number of fire-bomber conversions dwindled, military work alone was not enough to suffice. The changeover of aircraft types at Comox Station increased the number of Canadian-built machines that were to be returned to their manufacturer for overhauls.

Courtesy *Victoria Times Colonist*

At top of the page is the launching of *Hawaii Mars* (CF-LYL) at Patricia Bay on August 20, 1964; CF-LYK is in the distance. Immediately above is a classic newspaper shot of *Hawaii Mars* discharging a 30-ton drop.

In December 1967, Fairey Aviation closed the doors of its western plant at Victoria International Airport. March 1970 saw the closure of Fairey's east-coast plant, ending this famous company's operations in Canada.

Photo by Elwood White

*Philippine Mars* (foreground) and *Hawaii Mars* were based at Sproat Lake, Vancouver Island, supervised with loving care by W.F. (Bill) Waddington, Operations Manager and Chief Pilot. Forty years later, they are still incomparably effective.

In April 1974, Forest Industries Flying Tankers Ltd. published an informative brochure, detailing the specifications of the Mars Water Bombers and their achievements through 1973. Already they had dropped 14.8 million gallons of water in 2,893 drops at 170 fires. The numbers through 2004 would be stupendous.

## 26. The Flying Firemen: Canso Water Bombers ▪ 1965–1971

Fairey Aviation's technical success with water-pickup flying boats inspired Victoria's other major operation of this kind. In 1965, a company called The Flying Firemen, Ltd. was formed by Fairey's chief test pilot, Alex M. Davidson, in partnership with Bill Harrison, recently from Omineca Air Service of Burns Lake. Initially the partners leased two PBY Cansos (one American and one Canadian), which they later bought outright. Contracting to the B.C. Forest Service for the summer fire season of 1965, they lured out of retirement veteran bush pilot Sheldon Luck, who became The Flying Firemen's chief pilot. Those first two aircraft were converted into water bombers at Pat Bay by Fairey Aviation, and registered as No. 1 (CF-NTL) and No. 2 (CF-FFX); see photos below. No. 2 was a Super Canso that had received more powerful engines from a B-25 Mitchell. Each could carry over a thousand gallons.

Encouraged by their early success, Davidson and Harrison bought a third Canso in Mexico during the fall of 1966. Converted as Flying Firemen's No. 3 (CF-FFZ), it was ready for the fire season of 1967; see photo right.

That summer of 1967 was to prove disastrous, as a heat wave led to fires that consumed many thousands of acres of British Columbia forest. At the peak of the season, Sheldon Luck was flying No. 1 out of Prince George, while Alex Davidson was stationed at Patricia Bay with No. 2 and No. 3.

One Sunday afternoon in mid-July, the Provincial Government's Forest Service and the local public became alarmed about a large fire that had broken out suddenly on Skirt Mountain near Goldstream Park, just a short drive west of Victoria. Billows of smoke and even occasional jets of flame could be seen from higher parts of the city. The date was July 16, 1967.

The blaze was first reported around 4 p.m. It was said to be less than a third of a mile from the Trans-Canada Highway, where traffic had slowed to a crawl. The Flying Firemen were called in at 4:30 p.m. With his co-pilot R.T. "Paddy" Moore, a former Fleet Air Arm and Fairey Aviation test pilot, Alex Davidson made repeated passes over the fire in Flying Firemen No. 2 (CF-FFX), each time scooping up water from Finlayson Arm. About 6:30 p.m., as the fire still raged out of control, the aircraft was seen to come in very low on a hazardous pass over Skirt Mountain. Suddenly one wing-tip struck a high tree; the Super Canso disintegrated and plunged to the ground, killing both pilots instantly. The horror of the crash caused great dismay on southern Vancouver Island. The tragic event dominated local newspapers for days, as other water bombers were summoned to help conquer the stubborn fire—Sheldon Luck's Canso No. 1 from Prince George, and a Martin Mars from Sproat Lake.

Photo by Elwood White

Photo by Elwood White

Photo by Elwood White

CP wirephoto courtesy *Times Colonist*

(Right) Charred remains of Canso PBY water bomber CF-FFX on Skirt Mountain, sadly viewed by Forestry Department employee David Reid the next day.

The loss of two key men almost finished this fledgling company. Determined to keep going, however, Bill Harrison bought yet another surplus Canso in Mexico and converted it during the winter of 1967–68, ensuring that there would be three aircraft ready for the 1968 season. The new Canso had inherited the old No. 2 designation, with the new registration CF-FFY. On August 8, tragedy struck again. While fighting a particularly nasty fire near Shawnigan Lake, on the last drop of the day, the Canso flew into a ridge and crashed, killing pilot Tommy Swanson and engineer Tommy Worley. Aviation professionals had always acknowledged the dangers of aerial firefighting, but the shock of a second almost identical disaster was devastating.

Flying Firemen grimly decided to retire the ill-fated Number 2, and purchased another Canso that became No. 4 (CF-FFW). Perhaps too discouraged to continue, Bill Harrison then sold the firm, early in 1969.

Courtesy Bob Dyck

The new owner was a prominent businessman with an extensive background in aviation: R.L. (Bud) Rude, former owner-manager of Alaska-Central Airlines. Because Sheldon Luck had left the company in 1969 to tend his farm in northern B.C., veteran employee Frank J. Steven was promoted in 1970 to the position of Chief Pilot and Operations Manager. A former RCAF flying-boat captain and chief instructor of the Armed Forces flying-boat school on the Pacific Coast, he was ideally qualified for the position. Jack Ellard, also an ex-RCAF captain and more recently chief instructor and manager of the Victoria Flying Club, had become Flying Firemen's General Manager in December 1968.

The end of 1969 saw the purchase of two Cansos, both converted at the Flying Firemen's Pat Bay hangar by Sidney Aircraft Repair Ltd.—No. 5 (CF-IHN) and an unnumbered CF-IIW, to be based in the State of Washington. By July 1970, four machines were available for service in British Columbia—two based at Prince George and two at Pat Bay.

Under the expert guidance of Frank Steven, who himself had some 6500 hours of experience on Canso and Albatross aircraft, a pilot-training program was set up in 1969 that consisted of 65 hours of flying and 60 hours of ground-school, along with four days of combined operations at Kamloops with other companies in the same business. Even though all the Flying Firemen crew and captains were flying-boat veterans, each of them was put through the training period every spring in order to sharpen the finer points in the science of fire bombing. It was obvious that the company was determined to attract outstanding pilots and to move this hazardous profession beyond the bush-league level of flying.

Further additions for the fleet of "bird-dogs"—smaller aircraft that perform spotting and guidance duties for the water bombers—were a Grumman Widgeon (CF-PNT) and an Aero-Commander (CF-ONS). A charter employee of the company, Chief Engineer Angus Kieler (see photo, page 178), supervised a crew of four engineers and kept the fleet in top condition. By now The Flying Firemen owned East Camp Hangar 3, the former Department of Transport building. Having passed through its difficult years, this determined company was ideally poised for successful operations in the 1970s.

Photo by Elwood White

Sheldon Luck, legendary British Columbia bush pilot.

B.C. Forest Service 16437

Flying Firemen No. 3 (CF-FFZ) makes a drop on the Spectacle Lake (Malahat) fire of 1967.

# 27. Victoria Flying Services: Harbour Re-opened ▪ 1959–1971

Having suffered from acute boredom after his sale of B.C. Air Lines in 1956 (see Chapter 19), Bill Sylvester started Victoria Flying Services, Ltd., at Victoria International Airport three years later, on August 29, 1959. He was now 48. His partner and chief pilot was Joe Howroyd; associates included Stan Scurrah and E.A. Cox from B.C. Air Lines. In fact, Victoria Flying Services Ltd. now became Victoria agents for B.C. Air Lines.

Photo by Elwood White

Several times in the 1960s, Victoria Flying Services sponsored weekend Air Shows and Fly-ins at Victoria International, always drawing large crowds.

The new company was located in Hangar 15, on the west side of Victoria International Airport; it operated a Cessna sales agency, a charter service, a flying school, and a Class 3 harbour-to-harbour service from a float on Belleville Street, in front of the B.C. Legislature. That seaplane service, with two de Havilland Beavers, included stops in the Gulf Islands. At one point, VFS operated as many as twelve aircraft.

DND Photo 72022

Near the Victoria Flying Services hangar No. 15 in West Camp, at a Bill Sylvester Air Show, is Navy Lockheed T-33 No. 465 from VU-33 Squadron.

The greatest achievement of VFS was to re-open Victoria's Inner Harbour to commercial seaplane service—denied for 35 years. The federal and provincial governments seemed obliging; but Victoria City Council was not quick to agree.

On August 2, 1966, the City's Public Works Committee ruled in favour of air charter service from the harbour, and even then there was support for regular service. An Air Canada strike in November 1966 did much to boost VFS's cause; so in May 1968 a licence was granted for passenger service between Victoria and Vancouver, with stops at Gulf Islands points in between. By April 1969 there were three round trips daily (*Times* 1969.04.18: 38). Other airlines began to compete, with traffic growing by leaps and bounds over the next thirty years. The seaplane of choice was at first the de Havilland Beaver, followed soon afterwards by the splendid Twin Otter.

Bill Sylvester died in Victoria at the age of 65 on July 19, 1976. Only days before, he had sold his majority interest in Victoria Flying Services.

Elwood White Collection

In 1966, a justifiably proud Bill Sylvester (left) received the Robert S. Day Trophy for outstanding contributions to B.C. aviation. It is here presented by Art Sellers of Skyways, the 1965 recipient.

Courtesy Times Colonist: Victoria Times 1959.08.17

Victoria Inner Harbour, August 1959. An unidentified Grumman Goose is moored in front of the B.C. Legislature. It is probably a VIP charter, since Victoria Harbour was still officially closed for commercial air travel.

Courtesy Times Colonist: Victoria Times 1968.06.21

Victoria Inner Harbour, June 1968. Victoria Harbour has now been licensed for commercial seaplane traffic, and a Victoria Flying Services Cessna can be seen at an expanded Causeway float.

Photo by Elwood White

Victoria Inner Harbour, view to the northwest in summer 1971. Since 1968 the seaplane wharf has been enlarged, and there are now four aircraft moored at the terminal: Cessna CF-MOJ (Hard Metals Canada, Ltd.); two Victoria Flying Services de Havilland Beavers, one of which is CF-VFS; and an unidentified Grumman Goose.

Photo by Elwood White

Victoria Inner Harbour, view to the northeast in summer 1971. At the same wharf, on a different occasion, there are again four aircraft moored at the terminal. Behind a Fairchild 24 are two Victoria Flying Services de Havilland Beavers, one of which is CF-YYU, and an unidentified Cessna.

AVRO 504K TEAM AT PATRICIA BAY
FEBRUARY–MARCH 1967

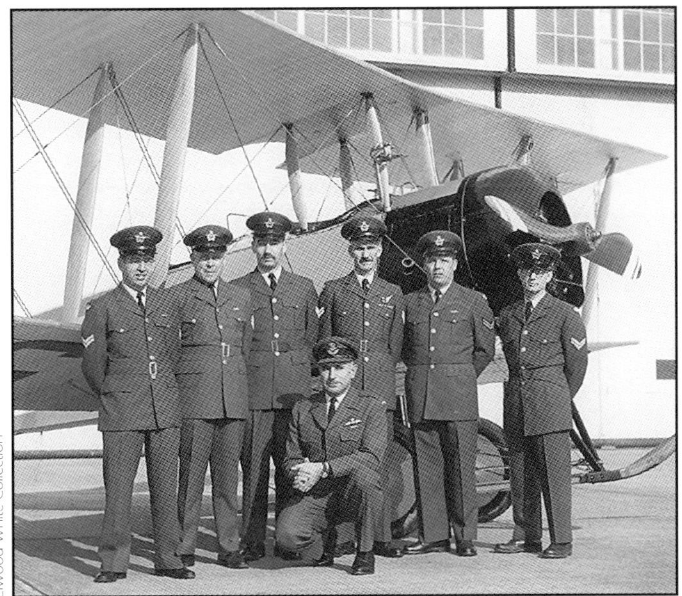

Elwood White Collection

This is a 1967 RCAF publicity picture of the Avro team.

Photos by Elwood White

Photos by Elwood White

During February and March 1967, a team of RCAF pilots took advantage of Victoria's balmy weather to rehearse a Centennial flying show to be presented across Canada later in the year with vintage Avro 504K biplanes, rebuilt for the occasion. In a *Colonist* feature on February 12, F/L Gordon Brown and F/L George Greff were shown performing aerobatic stunts in period flight suits. As the practice exercises continued, Elwood White took the pictures directly above of the team's aircraft at Victoria International Airport.

While enjoying an aerial joyride together in the course of the RCAF team's stay in Victoria, Elwood and Joe Howroyd discovered that one of the Avros had been required to make a forced landing on a Saanichton field. Help had already been summoned, and a rescue operation was in progress. As seen in the photographs at the left, Avro G-CYEI was eventually taken apart so that it might be transported back to Patricia Bay in sections. Joe Howroyd is clearly visible in both pictures.

## London-to-Victoria Air Race ▪ July 1 to July 7, 1971

One of the highly publicized events of the 1971 British Columbia Centennial Celebration—brainchild of Deputy Provincial Secretary L.J. (Lawrie) Wallace—was an air race between London, England, and Victoria. The race provides a symbolic conclusion to our first century of flight.

Sent on their way by Prime Minister Trudeau and Premier W.A.C. Bennett, the contestants took off from London on July 1, and arrived at Patricia Bay on July 7. Prizes were awarded within a complex system of classes and handicaps. Winner of the $50,000 grand prize was a 42-year-old West German pilot, Joseph H. Blumschein; Tim Phillips of Ireland was runner-up. Two Victoria pilots finished out of the money, but to great acclaim: Rick Cockburn, 25, and Claude Butler, 59, shared the Greater Victoria Chamber of Commerce "Man of the Year" award for 1971.

Photo by Elwood White

Winner Joe Blumschein flew this Texas-built Swearingen Merlin III.

Photo by Elwood White

Young Rick Cockburn of Sidney was admired for flying the race solo in this Harvard Mk.IV, CF-UZG. He quipped that his co-pilot was a spare gas tank.

Photo by Elwood White

John Howroyd Collection

Although neither of the Victoria entries finished within the top group, Claude Butler is clearly elated (left above) at having successfully completed the London–Victoria flight in his Aerostar 601, with co-pilot Harold Aasen of Delta. Their distinguished mechanic at Patricia Bay was Claude's own senior partner at Viking Air, Nils Christensen.

# 29. Looking Back Over a Century . . . With a Few Glances Ahead

Now that we have reached 1971, completing Victoria's first century of flight, we conclude with a short final chapter. It is time to express some general comments, tie up loose ends, and pay tribute to aviators we admire. Some of these are Victorians who made their mark on aviation far afield; others, who earned distinction elsewhere before choosing Victoria as their home, we salute as a group.

First, we acknowledge a fact that may be only too obvious to our readers. As this project took shape, we soon confronted a major problem: although we might be able to explore in some depth the history of local aviation up to about 1940, the period from 1941 to 1971 was far too complex to present in more than summary fashion. We have not tried even to summarize the growth of the Air Cadet movement, an important topic that we are not qualified to discuss. There are at least three other subjects, all admirably documented, that cry out for separate book-length treatment: Patricia Bay RCAF Station; Victoria International Airport (strangely neglected by Canadian aviation historians); and the dramatic saga of fire bomber development and use on southern Vancouver Island. On all three, we have barely scratched the surface.

We have seldom strayed beyond our stated terminal date of 1971, but we now make two important exceptions.

We express our admiration for recent progress by the Victoria Airport Authority, which has given Victoria International a welcome facelift for the 21st century. We look forward to other positive changes that are currently planned. The Airport has had a history of good leadership since the days of DOT managers K.J. Robinson (1948–1954) and Jack Knowland (1954–1977); under Victoria Airport Authority CEO Richard Paquette and board chair Linda Petch, it is in good hands in 2005.

We are also impressed with the swift advances made in some fifteen years by the British Columbia Aviation Museum, now broadly supported by many enthusiastic members. The Museum began with a dual interest in aircraft restoration and Pat Bay history. It is making splendid progress in documenting and creatively displaying the story of aviation in Greater Victoria, and is gradually assembling a truly outstanding collection of historic aircraft. A few are superior replicas, like the full-size model of William W. Gibson's Twin-plane. Most, however, are authentic vintage aircraft, lovingly and painstakingly restored by highly skilled craftsmen, all dedicated volunteers. Had it not been for the presence at Patricia Bay of such technically superior companies as Fairey Aviation and Viking Air, there would not exist today the deep pool of gifted professionals required for these very sophisticated restoration projects.

Photo by: Elwood White

Following the lead of Bill Sylvester's Victoria Flying Services, Norm Gold's AirWest Airlines of Vancouver pioneered Twin Otter flights out of Victoria Inner Harbour, soon followed by West Coast Air. Here in 1972 is AirWest D.H. Twin Otter CF-AWC. Beyond the cruise ship *West Star* are the Gulf Oil tanks in the Songhees Industrial Reserve. Not without some lively controversy, seaplane traffic would burgeon over the next thirty years.

One aspect of our research that has caused delight is the discovery of key individuals who recur at regular intervals throughout the story. This not only provides a sense of continuity, but it also reinforces the pleasant but mythical notion that nothing ever really changes in Victoria.

The Great War veterans from the Aerial League of Canada provide ideal examples. Gordon Cameron steps on stage in Chapter 5, assumes a major role with B.C. Airways in Chapter 8, and plays the elder statesman in Chapter 18 as an advocate for the Gordon Head Memorial Air Park. His Aerial League partner Jimmy Gray, a pioneer glider enthusiast, makes a cameo appearance in Chapter 9 aboard Maurice McGregor's Gypsy Moth CF-ADY, then later heeds the call of duty in World War II to serve as flight trainer for a brave new generation.

Courtesy of John D. Gray

In World War II, Squadron Leader James Gray had postings to Regina, Saskatoon, and Rivers, Manitoba. After the war he resumed his management of Marine Iron Works, the company that his father founded.

Born in Victoria on December 10, 1891, Jimmy Gray died in Victoria on October 3, 1983.

Purely by accident, we came across an item in the *Victoria Times* of January 23, 1946, that told of the return from California to Victoria of the fabled Harry Brown; he was visiting family members who all still lived on Green Street near Athletic Park, where he had grown up as a child. In the 1960s, his sidekick Bob Rideout retired to Vancouver after a wild and colourful life in Hollywood.

The same recurring principle applies to the boosters and backers of local aviation. Archie Wills is a prime case in point, again going back to the 1919 era of the Aerial League. Gaining seniority at the *Victoria Times*, Archie took the lead in persuading the Mears and Collyer duo to fly out of Lansdowne Field in 1928, and then Charles and Anne Morrow Lindbergh in 1931. As a veteran Victoria Alderman, he chaired the City's short-lived Airport Committee in the late 1930s, and sometimes attended ceremonial aviation events as Acting Mayor. Late in life and long into retirement, Archie wrote vivid accounts of Victoria's aviation highlights, regularly offering his lively articles to the rival *Daily Colonist*, a newspaper destined soon to merge with his own.

Prominent for decades among Victoria's movers and shakers were other shrewd air-minded men like R.H.B. Ker, William Straith, Frederick M. McGregor, and R.W. Mayhew. Many of them exerted their influence through the Victoria Chamber of Commerce. In 1937 Robert Mayhew became the city's virtually permanent Liberal Member of Parliament. At times he did not see eye to eye with all his influential friends, some of whom were not of his political persuasion. Still, R.W. Mayhew took second place to no one in his devotion to Victoria, and it gave him enormous pleasure in June 1943 to arrive at Pat Bay aboard the first truly transcontinental TCA flight.

Everyone in Victoria knew Fred McGregor, who had been a supporter of B.C. Airways long before his son Maurice decided—after calming his nervous parents—to become a pilot. Younger brother Raymond followed suit, and soon he and Maurice were both TCA captains. Fred could not possibly have been a prouder father.

Maurice McGregor Collection

March 1, 1939, Esquimalt Harbour. Fred McGregor (L) and Robert W. Mayhew (R) are about to board Canadian Airways' inaugural airmail flight to Vancouver, which will connect with the inaugural overnight airmail flight to Eastern Canada, piloted on its first leg by TCA Captain Maurice McGregor, Fred's elder son.

Perhaps Victoria has had some cause to feel neglected at times by the rest of Canada. Proud leader of British Columbia in pioneer decades, the city was gradually eclipsed by upstart Vancouver, a town that had even usurped our Island's name. Victoria seemed gradually to fade from national awareness. Before Newfoundland joined Confederation, Canada was routinely described as stretching "from Halifax to Vancouver"—a galling phrase. When the Canadian Broadcasting Corporation was established, B.C.'s Provincial Capital had no station for half a century. National air transport became another irritant. Although its island location cried out for service, Victoria waited a full six years for Trans-Canada Air Lines.

We have seen that the Victoria Chamber of Commerce often took a leading role in promoting the cause of civil aviation. Robert Mayhew himself had been Chamber president when Canadian Airways first came to town in 1931. On arriving aboard the first scheduled TCA flight in June 1943, he was greeted by Acting Mayor Archie Wills and a trio of Chamber of Commerce notables. Also present, as he had been in 1931, was George ("Follow the birds to Victoria") Warren, Director of the Victoria and Island Publicity Bureau.

Victoria then indulged in a gleeful civic celebration: self-congratulatory public events were prominently featured in both local newspapers. Naturally, there was some parochial satisfaction in the fact that Vancouver was no longer the western terminus of the national airline. But there was also a more serious belief that this event would transform Victoria's economy and lifestyle. In many ways, that prediction came true.

When Patricia Bay Airport was transferred to civilian management three years after World War II, Victorians would soon enjoy the best short-haul air service in Canada. Few other airports in the country could match its annual passenger count.

Although Victoria has grown and matured remarkably in the sixty-year period since R.W. Mayhew enjoyed this warm public reception, the city's role in national aviation may still be undervalued in Central Canada. We hope that our book may help to close that credibility gap.

Archie Wills Collection

Patricia Bay Airport: June 6, 1943. The first regularly scheduled TCA passenger plane to Victoria, a Lockheed 18 Lodestar, landed at 11:25 p.m.(see page 150). No one was more delighted than Victoria's Member of Parliament, Robert W. Mayhew (far left), who had flown all day from Ottawa. He was met that night by Acting Mayor Archie Wills (second from left) and other keen aviation boosters: (L–R) prominent businessman J.V. (Johnny) Johnson, George H. Harman, RAF veteran R.H.B. Ker, and George I. Warren (Victoria and Island Publicity Bureau). George Hamilton Harman, local Manager of the Bank of Montreal, was the most recent Chamber of Commerce President. While still in office, he died suddenly on the eve of V-E Day, May 7, 1945.

## POLITICAL BALANCING ACTS IN OTTAWA

The Honourable C.D. Howe took a lot of flak from Canadian critics who bristled at his brusque self-assurance, but we owe him a debt of gratitude for our excellent Victoria International Airport. There is a neat symmetry in the fact that a Liberal Cabinet Minister, C.D. Howe, opposed the Gordon Head Memorial Air Park in 1948, whereas a Conservative Cabinet Minister, Victoria's own Major-General George R. Pearkes, V.C., made it possible ten years later for all the acreage of both short-lived Gordon Head air fields to be purchased as the nucleus of a future campus for the University of Victoria. In retrospect, events seemed to turn out for the best.

N. A. YARROW,
PRESIDENT
L. A. DOBBIN,
MANAGER

TELEPHONE: SEY. 2831
NIGHT CALLS
BAYVIEW 7192
BAYVIEW 4882 L
VICTORIA OFFICE 2094
CABLES: "AIRCRAFT"

YARROW AIRCRAFT CORPORATION
LIMITED
CONTROLLING
DOMINION AIRWAYS LTD.

COMMERCIAL FLYING ACTIVITIES:
AERIAL PASSENGER, FREIGHT AND EXPRESS SERVICES
SIGHTSEEING TOURS AIR TAXI SERVICE
AERIAL FORESTRY PATROLS AND SURVEYS
AERIAL PHOTOGRAPHY AND SKETCHING

HEAD OFFICE:
1303 VANCOUVER BLOCK
VANCOUVER, BRITISH COLUMBIA
CANADA

GROUND FACILITIES:
AEROPLANE AND AERO ENGINE REPAIRS
AIRCRAFT ASSEMBLING AND SERVICING
AGENTS FOR
AIRCRAFT AND AERONAUTICAL EQUIPMENT
FLYING SCHOOL
DUAL AND SOLO FLYING GROUND COURSE

Courtesy Chris Yarrow

Norman A. Yarrow (1891–1964) was a profoundly important name in the history of Victoria ship-building— president and managing director of Yarrows Ltd. from 1914 to 1946, when he sold his shipyard to Burrard Dry Dock Co. Ltd. Son of Sir Alfred and Lady Yarrow, he emigrated to Canada in 1914, after his father had bought the British Columbia Marine Railways Co. Ltd. at Esquimalt. Norman Yarrow had a quiet passion for aviation.

In 1919–20 he strongly supported the Victoria Branch of the Aerial League of Canada, rebuilding the damaged *Pathfinder* and refitting *Pathfinder II* (page 32). In 1929, he acquired controlling interest of Vancouver's Dominion Airways Ltd., founded in 1927 and operated by Leonard A. Dobbin and his three brothers. The short-lived Yarrow Aircraft Corporation Ltd. was not a great success. Still, Fokker Universal CF-ABL, seen above in Esquimalt Harbour, created a sensation when it arrived in Victoria on May 13, 1929, as reported in the *Times* that evening.

Stinson Detroiter CF-AMD, bought in 1930 by Yarrow Aircraft Corp., is here moored at Victoria's B&K Wharf.

DH Moth CF-AAB (above) was bought in 1929 by Norman Yarrow, along with Moths CF-AAC and CF-ADY.

## Charles Melvin Price and His Restored 1930 Fleet Biplane, CF-AOD

Soon after the end of World War II, Mel Price bought a battered old Fleet 2 biplane, which he hoped to restore to its pristine condition. A professional mechanic with the Victoria Flying Club, he had the tools and the talent for the job; after countless hours of effort, the aircraft was as good as new.

Until a seaplane ban was imposed, Mel Price's Fleet biplane could often be seen at Elk Lake. Below left, CF-AOD is photographed in flight over the Saanich Peninsula, in April 1971; Elwood White's son Jamie is a passenger.

Mel flew the red-and-cream biplane regularly, either on floats or on wheels, and it became a very familiar sight at Patricia Bay.

By the late 1960s, CF-AOD had become the oldest registered aircraft still flying in Canada. Mel stated that he could still cruise at 95 mph on its 160hp Kinner five-cylinder engine. In June 1971, Price sold his Fleet for $10,000 to the B.C. Provincial Museum, later the RBCM (*Colonist* 1971.06.11: 40). By February 1972, it was suspended from the ceiling of that museum's new history display, once again on floats and now painted bright blue and yellow.

In recent years, Mel Price's vintage Fleet model 2, manufactured originally in Fort Erie, Ontario, has been on loan from the Royal British Columbia Museum to the British Columbia Aviation Museum, at Victoria International Airport.

All photos on this page by Elwood White

At left, Mel Price's 1930 Fleet 2 is seen in July 1970, a year before its sale to the B.C. Provincial Museum. Victoria Flying Club Hangar 4 is in the background. With the proceeds of his sale, Mel bought a gorgeous black-and-yellow 1929 Monocoach, registered as CF-AAT, which he proudly displays above at Butler-Howroyd Strip.

## Air Heroes of World War II

During the Great War of 1914–1918, perhaps a few dozen young aviators from Victoria served overseas with one of the British flying units. In the Second World War, the number of Victorians who joined the air force could be counted in the hundreds. For college students, in particular, the RCAF had a strong appeal, partly because of its glamour and partly because it seemed to offer a greater personal challenge. In the years of massive bombing raids over Germany, casualties became very heavy, and many brave young fliers did not live to see Canada again. We salute all these heroes as a group, and single out two famous and conspicuous pilots to represent their fellows. On remarkable but very different career paths, both these men continued in the armed forces after the war, and both returned in retirement years to their native city.

Born January 4, 1920, Reginald John Lane attended local schools, graduating from Victoria High in 1936. He joined the RCAF in 1940, was sent overseas on completion of pilot training, and received his commission in 1941. For Canadian bomber pilots in that stage of the war, completing two full tours of air combat duty was viewed as a normal maximum. Reg completed three tours, two of them in a Pathfinder Squadron that was required to act as an advance guard for every heavy bomber raid over Germany. Promoted to Group Captain in 1944, at 24, he was awarded the Distinguished Service Order, the Distinguished Flying Cross with bar, and a Mention in Dispatches. Reg married his wife Barbara in 1945, in England, and they eventually had four children.

His postwar career in the RCAF was characterized by a steady rise in rank and responsibility. On returning to Canada, he completed the RCAF staff college course at Toronto. In 1949 he became senior personnel staff officer of Air Transport Command, then located at Rockcliffe. Soon he was commanding officer of RCAF Station Edmonton.

After graduating in 1955 from the Imperial Defence College in London, England, Reginald Lane returned to RCAF headquarters in Ottawa as director of plans and programs, in 1958 becoming Chief of Plans and Intelligence. In June 1961 he was appointed Air Officer commanding Air Transport Command, with headquarters at Trenton, Ontario.

His career was plainly an uninterrupted succession of increasing responsibilities, a tribute to his remarkable talents. In January 1966 he became chief of staff, 1 Air Division headquarters, Metz, France. In August 1966 he was promoted to Air Vice-Marshal, becoming commander of 1 Air Division. When the Canadian Forces were reorganized on February 1, 1968, he assumed the rank of Major General. New duties followed until 1972, when he was promoted to Lieutenant General and appointed Deputy Commander-in-Chief at NORAD Headquarters, Colorado Springs. There he was named an Officer of the Legion of Merit by the President of the United States.

Lieutenant General Reginald J. Lane, DSO, DFC, CD

Canadian Armed Forces photograph

There can be few Canadian airmen who have compiled a more meritorious record of service. Even after he retired from the military in 1974, the remarkable pattern continued. During his long and fruitful years of retirement in Victoria, until his death at 83 on October 2, 2003, Reginald Lane was constantly active in rendering service to his country and his fellow citizens: Royal United Service Institute, Greater Victoria Art Gallery, Air Force Officers Association, Canadian Club, Twilight Homes Ltd., External Affairs Consultative Group on Arms Control and Disarmament, Canadian Institute for International Peace and Security—the list is endless.

Reg Lane was an uncommonly gifted and decent man, a wise and visionary leader who is remembered with praise and admiration by all who knew him.

Here were two men of roughly the same age, from the same Canadian city and a similar background, who served in the same global conflict. Each distinguished himself in combat and rose in postwar ranks. For whatever personal reasons, however, Vernon Crompton (Woody) Woodward was not destined to enjoy such continuing success and acclaim as Reginald Lane.

As war clouds formed in 1937–38, Vern Woodward was one of many young men who, frustrated by the high entry requirements of the RCAF, sought help from an unofficial Royal Air Force recruiter in Victoria, retired Royal Navy Captain Henry Seymour-Biggs. [For this story, see *The Biggs' Boys* (1995), by former RAF pilot Ken Stofer of Victoria.] So it was that 21-year-old "Woody" left Victoria in January 1938 as one of the "Biggs' Boys" to become a fighter pilot in the RAF. He received his commission in 1939, and was posted to No. 33 Squadron in Egypt. He would spend his combat years of the war (1940–1942) in the eastern Mediterranean, fighting aerial battles in the skies over Egypt, Libya, and Greece.

Though the details of these campaigns are complicated and the quoted numbers vary, his exploits are documented on the Internet and in a biography by Hugh Halliday titled *Woody: a Fighter Pilot's Album* (1987). By midsummer 1940, Woody had already shot down five Italian Fiat biplanes (CR32 and C42 models). By the end of 1941 he was a celebrity—at that point Canada's top-ranking aerial ace. His score was twenty-one enemy aircraft shot down, five probables, eleven damaged, plus two destroyed on the ground. Then given postings that were not in major combat zones, he saw his totals surpassed by George F. (Buzz) Beurling. Woody earned the Distinguished Flying Cross in 1941 and a Bar to the DFC in 1943.

Returning from the Middle East to England after the war, he accepted a permanent RAF commission in 1948, becoming C.O. of No. 19 Squadron, and he later served as Wing Commander in Germany and Malta. W/C Woodward retired from the RAF in January 1963, taking up residence in Australia. For the next four years he had substantial success with an air charter company.

In 1967 Vernon Woodward returned to Canada, setting up a business in Vancouver. That venture did not work out well, and Woody seemed to find it increasingly hard to cope. It appears, however, that he did find satisfaction in his final quiet career, an extended stint with the Corps of Commissionaires, posted often to Esquimalt Dockyard or to Victoria International Airport. After his death on May 26, 2000, a sensitive and balanced profile appeared in the *Times Colonist* (2000.07.02: B2). Though he was a brilliant pilot and a famous war hero, life had not been altogether kind to Vernon C. Woodward.

Canadian Armed Forces photograph

Wing Commander Vernon (Woody) Woodward, DFC & bar. A World War II ace, he flew Gloster Gladiator biplanes and Hawker Hurricanes in the Royal Air Force.

1936 Camosun Victoria High Archives

Reginald Lane and Vernon Woodward were Victoria High School classmates in the matriculation class of 1936—Reg in Division 4 and Vern in Division 2.

Photo by George Maude

This page illustrates two aircraft with strong Victoria connections that are now in Ottawa's Canada Aviation Museum, and a third vintage aircraft of prime museum quality.

At left is Bristol Bolingbroke 9892, which George Maude donated to the RCAF in 1964, having rescued it from demolition at Patricia Bay after World War II (see page 145). It is shown here in 1946 at Fulford Harbour, Saltspring, where George has just brought it by barge.

George Maude Collection

In the middle picture, George Maude sits in the cockpit of his P-40 Kittyhawk (see page 145), while son David Maude smiles approval. As we noted earlier, this classic fighter plane, still in Maude family possession, is considered one of the best preserved P-40s in the world. Its presence at Victoria International Airport stimulated the growth of what is now the B.C. Aviation Museum.

Aside from his passion for classic aircraft, George Maude is perhaps the foremost authority on the history of Victoria International Airport; he served at Patricia Bay RCAF Station from 1943 to 1945, before enjoying a 29-year career on site (1948–1977) with TCA and Air Canada. He has built up a unique collection of Pat Bay historic documents and artifacts.

Photo by Elwood White

In front of a Pat Bay hangar (left) is the famous RCMP Grumman Goose CF-MPG, long stationed in Victoria, now in the Canada Aviation Museum, Ottawa (CAM). A Grumman G-21, it is akin to the Turbo Goose modified in 1969 by Nils Christensen and Angus McKinnon (page 175). Entering RCAF service in 1944, it was operated by the RCMP from 1946 to 1994, "serving in continuous public use longer than any other Canadian aircraft" (CAM website). In 1995 it was donated to the Ottawa museum by the RCMP and the B.C. Government, after a public campaign that celebrated its distinguished service on the British Columbia coast.

## Nils Christensen, "Goose Doctor" and Eminent Flight Engineer

Proud son of Norway, Nils Christensen was flight engineer in 1961 aboard the first Mars water bomber (see Chapter 25). He had played a major role in developing that complex technology, and would become renowned for his skill in all aspects of aircraft conversion and maintenance. Founder of Viking Air at Victoria International Airport in 1970, he was inducted in 2004 as a charter member of the Canadian Aircraft Maintenance Engineers Hall of Fame (*Aviator Magazine*, March/April 2004, page 10).

He has had a remarkable life. Born in Oslo on August 15, 1921, he joined the Royal Norwegian Air Force in 1942, and was sent as an aircraft mechanic to Toronto's "Little Norway." He then saw overseas service in England as a flight mechanic and air gunner on Catalina flying boats. After the war, he served as chief engineer on the the personal aircraft of Norway's Crown Prince and future King.

After his return to Canada in 1951, Nils worked first as a powerplant mechanic for de Havilland and then became chief of maintenance at Sault Airways in Sault Ste. Marie. It was in 1956, at age 35, that he moved to British Columbia, to become chief of maintenance at the Victoria Flying Club. He would spend the rest of his working life at Patricia Bay, retiring evenually to Saltspring Island.

Receiving leave of absence from Victoria Flying Club in 1959, when appointed chief of maintenance for Flying Tankers, Nils took charge of 93 tons of spare parts for the Mars water bombers. He stayed on the consortium's payroll until 1966, when the Mars conversion program was ended, with the aircraft stationed at Sproat Lake.

Other challenges soon followed: from 1965 to 1967 he honed his skills at Fairey Aviation, and in 1967 he joined McKinnon Enterprises as shop foreman, when that company was awarded a B.C. Government contract. It was Nils who converted the B.C. Government's Grumman Widgeon CF-GPJ ("Gaglardi's Pride and Joy") and the Grumman Turbo-Goose (Chapter 24). He became so expert in adapting and rebuilding Grumman amphibians that he has been nicknamed "The Goose Doctor."

Nils Christensen took a bold leap in 1970 by founding his own company, Viking Air Limited, in partnership with Courtney Griffiths and Claude Butler. Acquiring state-of-the-art equipment from McKinnon Enterprises, Viking Air soon established its own reputation for outstanding aeronautical engineering. Thanks to Nils Christensen, his company obtained the de Havilland parts manufacturing rights for the Beaver, Otter, and Turbo Beaver aircraft.

Viking Air's payroll grew to about fifty skilled workers under Nils' leadership, from 1970 to 1987. In the 21st century it is still one of the airport's major employers.

In this 1986 photograph, Viking Air occupies the middle (No. 2) East Camp Hangar at Victoria International Airport.

Photos from Nils Christensen Collection

Mention was made earlier of aviators who achieved distinction elsewhere in Canada or abroad before moving to Victoria in retirement. A man familiar to our readers, representing both military and civil aviation, is Major Donald R. MacLaren (1893–1989), highly decorated Great War ace and Hall of Famer who became a key figure in both Canadian Airways and Trans-Canada Air Lines. Of course, he had been a frequent visitor to Victoria for many decades before becoming a resident late in life.

In Canadian commercial aviation, another member of Canada's Hall of Fame was Captain William John (Jack) Sanderson, who was born in Lakewood, Ohio on November 24, 1898, and who lived for years in Victoria before his death on January 22, 1984. He was an RFC veteran who had honed his flying skills when he returned after World War I to his home town of London, Ontario. Jack Sanderson met Major Reuben H. Fleet in 1929, and in 1930 founded Fleet Aircraft of Canada, Limited. Fleets from Fort Erie, Ontario were for years among the most popular private airplanes in Canada; see the Fleet Canucks below, and earlier models on pages 82–83.

Donald R. MacLaren and W.J. Sanderson are notable examples of local retirees from the older generation of aviators. If asked to select one representative member of the World War II generation who eventually retired to Victoria, we could choose no more worthy candidate than the late Squadron Leader T.P.M. (Mike) Cooper-Slipper, a decorated RAF fighter pilot who has also become a member of Canada's Aviation Hall of Fame.

Born in Kinver, Staffordshire on January 11, 1921, Mike Cooper-Slipper became a well-known resident of Victoria—with his wife Rita—for eighteen years prior to his death on February 23, 2004. He was not only a celebrated war hero; he had also made unique and important contributions to Canadian civil aviation.

For his exploits in the Battle of Britain he earned a DFC at the precocious age of 19, flying Hurricanes for RAF No. 605 Squadron. Posted to Singapore and North Africa, he ended the war with nine combat victories.

After emigrating to Canada with Rita in 1947, Mike enjoyed a long and eminent career as a senior test pilot with Avro Canada and as Chief Test Pilot with Orenda Engines. Mike was most notably involved with the flight test development of the Avro Jetliner and the Iroquois engine designed for use in the Avro Arrow. Sadly both the Jetliner and Arrow were destined for cancellation. Mike worked in aircraft sales from 1960 to 1972 and for the Ontario Government until 1986 as a Development Officer for the Ministry of Industry and Trade.

Courtesy Chris Cooper-Slipper

Avro Canada test pilot Mike Cooper-Slipper wears a high-altitude pressure suit while climbing into an Avro (Canada) CF-100 Canuck, circa 1952.

Photo by Elwood White

In 1961, Victoria Flying Club Fleet Canucks CF-DQE, CF-DQA, and CF-DQR are lined up facing the Club's Hangar 3 (out of sight to the right). Jutting out beyond Fairey Aviation's Hangar 2 is the wing of *Philippine Mars* CF-LYK, now converted and freshly painted, perhaps ready for launching.

Maybe it is ironic that a world-famous Victoria aviator is known mainly for his repeated failures to fly across the Pacific. In Chapter 8 (p. 66), we observed that Albert Harold Bromley, born and raised in Victoria, was the central figure in a widely publicized venture that became an embarrassing fiasco. Bromley's Lockheed Explorer 4 (NR-856H), named *City of Tacoma,* crashed during takeoff on July 29, 1929. Two similar attempts soon afterwards to fly solo across the Pacific from Tacoma were also non-starters. Lougheed test pilot Ben Catlin was killed on the second try.

Archie Wills Collection

On September 15, 1930, pilot Lieut. Harold Bromley and navigator Harold Gatty took off from a wide sandy beach north of Tokyo, resolved to fly non-stop to Tacoma. Their aircraft, again named *City of Tacoma,* was an Emsco B-3A (NR 153W); they had brought it by ship to Japan (see photo, right). Gatty was a clever young Australian who had moved recently to America and was already attracting the attention and respect of Charles Lindbergh. In 1931, he would fly around the world with Wiley Post.

The two aviators were counting on a brisk tailwind, which never materialized. To compound the problem, escaping fumes had caused them both to suffer carbon monoxide poisoning, Bromley to the extent that his behaviour became wild and irrational. It was only by firmness and good fortune, it seems, that Gatty prevailed on him to abort the attempt and to limp back to Japan. Even so, Gatty would say of Bromley years later, "He never made the big time, but he was a magnificent pilot." See "Harold Gatty: Prince of Navigators," on the reliable website *The History Net* (historynet.com), which provides a detailed account of this almost disastrous flight.

The offbeat website *Goodbye* reported Harold Bromley's death in 1997 at age 99, stating that he taught Amelia Earhart to fly. If so, that is surely a better claim to fame than a series of fizzled would-be ocean crossings.

## A LOCAL PILOT REDISCOVERED

When this book was almost finished, out of the blue we learned the full story of yet another Victoria World War I pilot who had moved south to the U.S.A. Lawrence Louis Grant appears on pages 22–23 as Vice-President of the Aerial League of Canada, Victoria Branch. Information provided by his grandson, Donald Ward of Vacaville, California, revealed that Louis was born in 1895 in Union (Bay), B.C., where his father was a sawmill owner. His parents, Robert and Barbara Grant, were both pioneers from the Maritimes who were married in 1880 in Victoria before moving up-island. They returned to the city about 1910, when Louis was 15; in 1930, so we are told, they were the first couple to have been married in Victoria and to celebrate their golden wedding in the same city (*Colonist* 1941.02.01: 18).

As Aerial League VP, Louis was best known for his supporting role in the Rideout and Brown flight to Seattle in May 1919, and for his pioneer flight with Jack Clemence to Port Angeles on July 11, 1919. Soon after, he moved to Los Angeles, where he married and raised a family of three daughters. His 1972 obituary shows him posing beside Fred Astaire in front of a JN-4 used in the 1939 movie, *The Story of Vernon and Irene Castle.* By virtue of his service in the Royal Flying Corps, he had been a valuable technical advisor to RKO Pictures.

Archie Wills Collection

Harold Bromley (L) and Harold Gatty (R), pose with their Emsco aboard ship as they arrive at Yokohama.

## Victoria's Maurice McGregor: A Pilot Soaring to New Heights

Unlike the ill-starred Harold Bromley, Maurice McGregor was one Victoria aviator who quietly achieved all his goals and aspirations, enjoying continual success. In Chapter 8 we introduced him as a Canadian flyer whose career covered the whole spectrum of aeronautical experience: stunt flyer, instructor, pioneer bush pilot, airline captain, senior TCA and CPAL executive, aviation diplomat, international consultant, and airline owner. We conclude this book with a short tribute to Frederick Maurice McGregor (1911–1995), a man whose varied and eventful career could easily fill a substantial book on its own.

From his inaugural 1937 flight to Seattle through 1941, Maurice was one of TCA's Senior Captains, specializing in the Vancouver–Lethbridge night run over the Rockies, but also flying major routes in eastern Canada. In 1940 he served also as Assistant Superintendent, Flight Operations, for TCA's Western Division. From 1941 to 1945, he was Operations Superintendent for TCA's Eastern Division, with Headquarters in Toronto. From 1945 to 1946 he was Operations Manager of the Canadian Government Trans-Atlantic Air Service, responsible for converting the wartime Lancastrian service to a postwar commercial system. His last position with TCA, from 1946 to 1952, was Operations Manager, Trans-Canada Air Atlantic Limited, developing passenger service to the U.K., Eire, Bermuda, and the Caribbean.

Leaving TCA in 1952, Maurice McGregor joined Canadian Pacific Airlines, reporting to CPAL President Grant McConachie as Director of Development. This stage in Maurice's career can best be described as aviation diplomacy. After negotiating complex bilateral agreements with Peru and Mexico, adding those countries to the CPAL system, he became Manager of Overseas Lines, responsible for all north and south Pacific services.

As a senior TCA executive, Maurice could hobnob with celebrities like World Figure Skating Champion Barbara Ann Scott, welcomed to Montreal on March 10, 1947.

In 1954, through the International Civil Aviation Organization, Maurice was asked, with the approval of the Rt. Hon. C.D. Howe, to become General Manager of the proposed new Pakistan International Airlines (PIA). Granted leave by CPAL, he accepted this new challenge.

All photos from F.M. McGregor Collection

During their stay in Pakistan, Maurice and Joyce McGregor visited many regions of that historic country to assess tourism potential. Here we see them (left) at the Khyber Pass, where Maurice (right) poses with local boys.

For the early history of B.C. Air Lines, see Chapter 19, pages 158–160.

Here is Maurice relaxing en route from Karachi to London, after the successful completion of his Pakistan assignment. Within two years, he had helped the government of that country establish a modern and efficient international airline (Schofield and Corley-Smith 1992: "From Lansdowne to Islamabad").

Returning to B.C., he soon found himself co-owner and president of B.C. Air Lines, a company that grew and prospered under his supervision until its sale in 1960.

Maurice and Joyce McGregor would enjoy thirty-five more years in his native Victoria, a period when Maurice devoted himself primarily to other business ventures, while maintaining his pilot's licence and his keen interest in all aspects of aviation. Highlights included TCA's 40th Anniversay celebration in Seattle on September 1, 1977, and a triumphant return visit to Pakistan, at that country's invitation, in 1992. The young boy from St. Michael's School had truly achieved the life of his dreams.

At Seattle's Boeing Field for Air Canada's 40th Anniversary celebration on September 1, 1977, Maurice McGregor examines the vintage Boeing P-12 pursuit plane seen also in the photograph below.

Seattle's Boeing Field: Air Canada's 40th Anniversary celebration on September 1, 1977. Boeing B-727 (C-GAAJ) replaces 1937'S Lockheed 10A Electra. (L–R) Air Canada Board Chairman Pierre Taschereau, Bill Boeing Jr., Maurice McGregor (First Officer to Billy Wells on September 1, 1937).

At the official opening of this airfield almost half a century earlier, on July 26, 1928, Bill Boeing Sr. had welcomed a B.C. Airways delegation from Victoria. He pronounced their Ford-Stout Trimotor airliner to be "as safe as a church"—an inauspicious comparison, as events would prove.

All photos from F.M. McGregor Collection

## Bibliography

The authors wish to thank and acknowledge their indebtedness to the Canadian Aviation Historical Society, whose journal has done so much to foster research. In particular, a book such as ours would have been almost impossible without the two collected volumes of the *Canadian Civil Aircraft Register,* compiled for CAHS by John R. Ellis.

**Abbreviations**

| | | | |
|---|---|---|---|
| AAHS | American Aviation Historical Society | BCAM | British Columbia Aviation Museum (Sidney, BC) |
| CAHS | Canadian Aviation Historical Society | CMFT | Canadian Museum of Flight and Transportation (Langley, BC) |
| CCAR | Canadian Civil Aircraft Register | | |


Arbuckle, J. Graeme. 1987. *Badges of the Canadian Navy.* Halifax: Nimbus Publishing.

Bain, Donald M. 1987. *Canadian Pacific Air Lines: Its History and Aircraft.* Calgary: Kishorn Publications.

Ballentine, Gordon with Paul Stoddart. 1990. "They Don't Make 'Em Anymore." *Raincoast Chronicles 12: Stories and History of the BC Coast,* ed. Howard White: 64–76. Madeira Park, BC: Harbour Publishing.

Bashow, David L. 2000. *Knights of the Air: Canadian Fighter Pilots in the First World War.* Toronto: McArthur.

Blatherwick, John. 1989. *A History of Airlines in Canada.* Toronto: Unitrade Press.

Bowers, Peter M. 1966. *Boeing Aircraft Since 1916.* London: Putnam.

Brown, Jim. 1994. "Mr. Boeing & his Mechanician." *BC Aviator* 4.2: 7–9, 20–21.

————. 1996. *Hubbard: The Forgotten Boeing Aviator.* Seattle: Peanut Butter Publishing.

————. 2002. "When Victoria had a Chinese flying school." *Times Colonist Monitor* 2002.01.13: D12.

Bungey, Lloyd M. 1992. *Pioneering Aviation in the West: As told by the pioneers.* Surrey, B.C. and Blaine, Washington: Hancock House Publishers. (CMFT)

Cavin, Desmond J. 1989. "Panhandle Pilot," *Saanich: An Illustrated History.* The Corporation of the District of Saanich.

*CCAR 1920–1928. The Canadian Civil Aircraft Register, G-CAAA to G-CAXP.* Compiled by John R. Ellis and originally issued serially in *CAHS Journal.*

*CCAR 1929–1945. The Canadian Civil Aircraft Register [CF-AAA- ].* Compiled by John R. Ellis and originally issued serially in *CAHS Journal.* Published by CAHS.

CMFT. 1983. "VU-33 Squadron: Naval Aviation on the West Coast," *CMFT Museum Newsletter* No. 21: 24–25.

Corley-Smith, Peter. 1989. *Barnstorming to Bush Flying: British Columbia's Aviation Pioneers 1910–1930.* Victoria: Sono Nis Press.

————. 1993. *Bush Flying to Blind Flying: British Columbia's Aviation Pioneers 1930–1940.* Victoria: Sono Nis Press.

Corley-Smith, Peter and David N. Parker. 1998. *Helicopters: The British Columbia Story.* Victoria: Sono Nis Press. (Royal British Columbia Museum)

Douglas, W.A.B. 1986. *The Creation of a National Air Force. The Official History of the Royal Canadian Air Force, Volume II.* Toronto: University of Toronto Press.

Drew, George A. 1930. *Canada's Fighting Airmen.* Toronto: MacLean Publishing.

Duffy, Dennis and Carol Crane, eds. 1980. *Magnificent Distances: Early Aviation in British Columbia, 1910–1940.* Sound Heritage Series 28. Victoria: Provincial Archives of British Columbia.

Ellis, Frank H. 1944. "William Wallace Gibson: a Canadian Pioneer of the Air." *British Columbia Historical Quarterly* 8.2 (April 1944): 93–105.

————. 1961. *Canada's Flying Heritage.* 2nd ed. Toronto: University of Toronto Press.

Ellis, John R. *see CCAR.*

Feast, A.M. 1990. "The Gentle Giants of Sproat Lake," *Raincoast Chronicles 12: Stories and History of the BC Coast,* ed. Howard White: 8–22. Madeira Park, BC: Harbour Publishing.

Fuller, G.A., J.A. Griffin, K.M. Molson. 1983. *125 Years of Canadian Aeronautics: a Chronology 1840–1965.* Willowdale, Ontario: CAHS.

Hatch, F.J. 1981. "Spirit of Little Norway," *High Flight* 1.5: 171–180.

Jupp, Ursula. 1975. *From Cordwood to Campus in Gordon Head 1852–1959.* Victoria: Morriss Printing.

Kealy, J.D.F. and E.C. Russell. 1967. *A History of Canadian Naval Aviation,* 1918–1962. Ottawa: Queen's Printer.

Kostenuk, Samuel and John Griffin. 1977. *RCAF: Squadron Histories and Aircraft, 1924–1968.* Toronto and Sarasota: Samuel Stevens/Hakkert & Company. (National Museum of Man, National Museums of Canada)

Larkins, William T. 1957. *The Ford Story: A Pictorial History of the Ford Tri-motor 1927–1957.* Wichita, Kansas: Robert L. Longo Company.

Maude, David. 1995. "Little Norway—West!" *Flypaper* 1.3, unpaginated. Newsletter of the West Coast Museum of Flying, now British Columbia Aviation Museum.

McGregor, Maurice, as told to Paul Stoddart. 1991. "Flying for the Sake of Flying," *Raincoast Chronicles 13: Stories and History of the BC Coast,* ed. Howard White: 64–66. Madeira Park, BC: Harbour Publishing.

McKibben, Sherry. 1979. "Victoria Flying Club: fifty-two years of aviation history," *Colonist Islander* 1979.08.05: 4–5.

Milberry, Larry, ed. 1984. *Sixty Years: The RCAF and CF Air Command 1924-1984.* Toronto: CANAV Books.

Molson, K.M. 1974. *Pioneering in Canadian Air Transport.* Altona, Manitoba: D.W. Friesen.

————. 1980. "The First 500 Canadian Civil Pilots." *CAHS Journal* 18.1: 19–27; 18.2: 61–62.

Pigott, Peter. 1996. *Gateways: Airports of Canada.* Lawrencetown Beach, N.S: Pottersfield Press.

————. 2001. *National Treasure: The History of Trans Canada Airlines.* Madeira Park, BC: Harbour Publishing.

Ruotsala, James A. 1980. "Lockheed Vegas in Southeastern Alaska," *AAHS Journal* 25.1: 2–9.

————. 1997. *Pilots of the Panhandle: Aviation in Southeast Alaska, The Early Years 1920–1935.* Juneau: Seadrome Press.

Satterfield, Archie, with photos by Lloyd Jarman. 1969. *Alaska Bush Pilots: In the Float Country.* Seattle: Superior Publishing Company.

Schaffter, John. 1989. "Old Boy glad he dared to fly," *Times-Colonist Islander* 1989.03.26: M2

————. 1991. "Maurice knows joy of flying," *Times-Colonist Islander* 1991.05.12: M2

Schofield, Jack. 2004. *No Numbered Runways: Floatplane Pioneers of the West Coast.* Winlaw, BC: Sono Nis.

Schofield, Jack and Peter Corley-Smith. 1992. "From Lansdowne to Islamabad," *BC Aviator* 2.1: 7–11.

Turner, Robert D. 1977. *The Pacific Princesses.* Victoria: Sono Nis Press.

Vincent, Carl. 1980A. "Vancouver's Weekend Warriors: the story of 111 (CAC) Squadron, 1932–1941," *High Flight* 1.1: 25–34.

————. 1980B. "The First Thunderbirds: the story of 111 (F) Squadron, 1941–1945," *High Flight* 1.2: 70–78.

————. 1982. "Norse Pacific Saga." *High Flight* 2.5: 177–178.

Weicht, Christopher. 1997. *Jericho Beach and the West Coast Flying Boat Stations.* Chemainus.

White, Elwood. 1965. "The Saga of Junkers CF-AMX," *CAHS Journal* 3.1: 13–14.

————. 1995. "Alaska–Washington Happened in B.C," *West Coast Aviator* 4.5: 6–10, 20–21.

Wills, Archie H. 1969. "Lansdowne Airport: Victoria's Hope for Air Supremacy," *Colonist Islander* 1969.02.16: 4–5, 11.

————. 1978. "From 'Jennies' to 'Jumbos'," *Colonist Islander* 1978.08.27: 12–13, 15.

Wilson, A.H. 1965. "Reminiscences of a West Coast Pilot," *CAHS Journal* 3.3: 67–70.

Windsor, John. 1964. *Nowhere Else to Go* [a biography of Jerry Gosley]. Sidney, BC: Gray's Publishing.

Worthylake, Mary M. 1988. *Up in the Air.* Bend, Oregon: Maverick Publications.

Archie Wills Collection

Archie Wills treasured this photograph of B.C. Airways' Ford Trimotor arriving at Lansdowne Field, July 23, 1928.

# General Index

# Index of Aircraft and Registration Data